MW00611352

About Island Press

Since 1984, the nonprofit organization Island Press has been stimulating, shaping, and communicating ideas that are essential for solving environmental problems worldwide. With more than 1,000 titles in print and some 30 new releases each year, we are the nation's leading publisher on environmental issues. We identify innovative thinkers and emerging trends in the environmental field. We work with world-renowned experts and authors to develop cross-disciplinary solutions to environmental challenges.

Island Press designs and executes educational campaigns in conjunction with our authors to communicate their critical messages in print, in person, and online using the latest technologies, innovative programs, and the media. Our goal is to reach targeted audiences—scientists, policymakers, environmental advocates, urban planners, the media, and concerned citizens—with information that can be used to create the framework for long-term ecological health and human well-being.

Island Press gratefully acknowledges major support from The Bobolink Foundation, Caldera Foundation, The Curtis and Edith Munson Foundation, The Forrest C. and Frances H. Lattner Foundation, The JPB Foundation, The Kresge Foundation, The Summit Charitable Foundation, Inc., and many other generous organizations and individuals.

Generous support for the publication of this book was provided by Margot and John Ernst.

The opinions expressed in this book are those of the author(s) and do not necessarily reflect the views of our supporters.

Nourished Planet

Sustainability in the Global Food System

Edited by Danielle Nierenberg, Food Tank

Laurie Fisher, Brian Frederick, and Michael Peñuelas

Washington | Covelo | London

© 2018, Barilla Center for Food & Nutrition

All rights reserved under International and Pan-American Copyright Conventions. No part of this book may be reproduced in any form or by any means without permission in writing from the publisher: Island Press, Suite 650, 2000 M Street, NW, Washington, DC 20036

Island Press is a trademark of the Center for Resource Economics.

Library of Congress Control Number: 2017960090

All Island Press books are printed on environmentally responsible materials.

Manufactured in the United States of America
10 9 8 7 6 5 4 3 2 1

Keywords: regenerative agriculture, food access, food deserts, soil health, pesticides, water conservation, sustainability of well-being, the double pyramid, supply chain, nutrition, culinary traditions

Barilla Center for Food & Nutrition

BCFN Advisory Board

Barbara Buchner, Ellen Gustafson, Danielle Nierenberg, Livia Pomodoro, Gabriele Riccardi, Camillo Ricordi, Riccardo Valentini, and Stefano Zamagni

Food Tank Staff

Danielle Nierenberg, Bernard Pollack, Vanesa Botero-Lowry, McKenna Hayes, Emily Payne, Michael Peñuelas, and Brian Frederick

Nourished Planet is based on and has content from *Eating Planet Food and Sustainability: Building Our Future,* written by the Barilla Center for Food & Nutrition (Edizioni Ambiente, 2016). To discover more: www.barillacfn.com

Contents

FOREWORD

Valuing the True Cost of Food

Pavan Sukhdev and Alexander Müller

Food is more than sustenance. It is an integral part of the economic and sociocultural ecosystem in which we all live and work.

Unfortunately, there are significant externalities, both positive and negative, that prevent many of us from understanding the true cost of the food in our bowls or on our plates. Indeed, many of the largest impacts on the health of humans, ecosystems, agricultural lands, waters, and seas arising from various types of agricultural and food systems are economically invisible. As a result, they do not get the attention they deserve from decision makers in policy and business—or from eaters.

There is an urgent need to evaluate all significant externalities of eco-agri-food systems, to better inform decision makers in governments, businesses, and farms. And there is a need to evaluate the eco-agri-food system complex as a whole, not as a set of silos. The book you hold in your hands right now, *Nourished Planet*, represents a significant step in the right direction for enlightening policymakers, businesses, and society at large about the many dimensions of our eco-agri-food systems. This book focuses on not only the problems—of hunger, obesity, climate change, and poor nutrition—but also on the solutions.

As part of our work with TEEBAgriFood, a global endeavor to understand the size and scale of externalities along the value chain in different types of eco-agri-food systems, we have learned that we must understand not only the negative impacts of food production but also the diverse benefits that come from growing, transporting, distributing, and eating food in environmentally, economically, and socially sustainable ways.

Many of the benefits and costs attributable to agriculture are invisible; they are not traded in global markets and are not part of a price consumers pay at the checkout counter. But they do affect human well-being and the well-being of the planet. All these impacts, visible and invisible, must be assembled and evaluated through a universal framework in order to provide analytical consistency and comparability across systems, across policies, and across business strategies—in other words, across every aspect of our lives. And although this is no small feat, it can and should be done.

One exciting outcome of TEEBAgriFood's work is our analysis of rice production systems. Worldwide, about 80 million hectares of irrigated lowland rice provides 75 percent of global rice production. This predominant type of rice system receives about 40 percent of the world's total irrigation water and 30 percent of the world's freshwater resources withdrawn from the natural cycle. Our study compared the System of Rice Intensification (SRI) with conventional production methods and found that in Senegal, the cost of water consumption under conventional systems was significantly higher than it was under SRI. Switching to SRI methods, the study said, could reduce water consumption–related health and environmental costs in Senegal by about US$11 million a year. At the same time, the rice-producing community would gain about US$17 million through yield increases. This sort of win–win scenario has far-reaching benefits that can't be always be measured in real time, including multigenerational benefits as children see the value of this type of farming in their daily lives.

Top-down solutions rarely have the same economic or environmental impact that those coming from the ground up can have, and *Nourished Planet* makes it clear that discovering solutions means listening to communities and understanding their needs and wants.

In addition, *Nourished Planet* highlights the importance of reviving the fundamental aspects of eating that are most focused on the bond between food, the individual, and her community. At the environmental and ecological level, this will be about protecting local crop varieties, preserving biological diversity. At the social level, this will be about transferring the culinary expertise and know-how about preparing and serving foods in unique and culturally enriching ways, returning to a healthy relationship with the land and with the raw material by focusing on the excellence in quality of the ingredients, recovering age-old flavors, perhaps even making contemporary variants, and thus leading to the preservation of the best of the local culinary tradition.

Together with economic analysis of these challenges and solutions—including the work of TEEBAgriFood—*Nourished Planet* will be among the important works that contribute to a much better and holistic understanding of our food challenges. It will help create better and lasting food solutions for all—for the poor, for development, for the planet, and for society and culture—for generations to come.

Pavan Sukdev is the UNEP Global Ambassador and, on behalf of this UN agency, has led the project on *The Economics of Ecosystems and Biodiversity* (TEEB), commissioned to the United Nations by the European Commission and the German government. He is the founder–CEO of GIST Advisory, an environmental consulting firm helping governments and corporations manage their impact on natural and human capital.

Alexander Müller, a graduate sociologist, is directing a global study for the UN Environment Programme (UNEP) on "The Economics of Ecosystems and Biodiversity for Agriculture and Food" and managing the TMG: Töpfer, Müller, Gaßner GmbH, ThinkTank for Sustainability. From 2006 to 2013, he was assistant director-general of the Natural Resources Management and Environment Department at the Food and Agriculture Organization of the United Nations (FAO).

Preface

by Guido Barilla

The Barilla Center for Food & Nutrition (BCFN) was set up in 2009 as a research center to deepen our understanding of the complex global issues related to food, agriculture, and nutrition.

Thanks to the BCFN's multidisciplinary studies, we soon became aware of the paradoxes in the world's food system. People are starving, but obesity levels are rising and we still waste a huge amount of food. People go to bed hungry, but we increasingly use crops to feed animals and cars. Meanwhile, unsustainable farming pushes the environment to the limit.

The BCFN therefore started to develop ways to respond to these problems, in the context of the United Nations Sustainable Development Goals. The breakthrough came with a model that highlights the strong impact our food choices have on the environment. The message delivered by the BCFN's **double food and environmental pyramid** is simple: If we eat well, by reducing our meat consumption and focusing on a more plant-based diet full of whole grains and good fats such as olive oil, we can improve our own health and the health of the planet.

In 2014, the BCFN Foundation developed the **Milan Protocol** to raise awareness among institutions and the wider public on the need to tackle the world's food paradoxes, proposing ways to promote healthy lifestyles, encourage more sustainable agriculture, and reduce food waste.

The **Milan Protocol** inspired the **Milan Charter,** a global agreement to guarantee healthy, safe, and sufficient food for all, which the Italian government presented to Secretary-General of the United Nations Ban Ki-moon during Expo2015 (a global gathering of 145 countries in Milan around the theme "Feeding the Planet, Energy for Life").

The BCFN also encouraged young people to have a strong leadership role as part of the Expo2015 experience, with the creation of the Youth Manifesto, which calls for new policies to end the current paradoxes ruling food production, distribution, and consumption.

Youth have always played an important role at the BCFN. In addition to having several young researchers on staff, we have made a commitment to cultivate the next generation of food system leaders through the BCFN **Young Earth Solutions** (BCFN YES!) contest. To foster networking among contestants, the BCFN also started the BCFN Alumni Association.

The joint efforts of the BCFN's Scientific Committee and the Alumni have recently yielded two important projects aimed at helping to fix the world's broken food system.

The **Food Sustainability Index** (FSI), developed in 2016 with the Economist Intelligence Unit, highlights gaps and provides replicable examples of practices and policies to promote sustainable food systems. The FSI compares the work of countries using defined indicators to reveal the frontrunners—and how they got there—against three pillars: sustainable agriculture, nutritional challenges, and food loss and waste. The FSI is designed to be a reference point for policymakers and experts to orient their action and a tool to educate members of the public and inspire behavior changes for the good of our health and of our planet.

Born as a result of the Youth Manifesto, the **Food Sustainability Media Award** is the first international prize of its kind to reward excellence in journalism about food and sustainability. By informing and shedding light on today's food paradoxes, we believe the media at large can engage consumers so that in turn they can contribute to the creation of a more equitable and sustainable future, starting from their food choices. It was launched in partnership with the Thomson Reuters Foundation, and the first prizes to professional journalists and emerging talents in the media were awarded at the BCFN's international forum in December 2017.

This edition of *Nourished Planet* is meant as a guide for youth who care about where their food comes from, who produced it, how it was grown, and the sorts of policies and practices that will not only feed future generations but nourish them as well.

Guido Barilla is chairman of the Barilla Center for Food & Nutrition Foundation.

Preface

by Danielle Nierenberg

As I write this, I'm watching the devastating images of Hurricane Harvey in Texas, Florida, and Puerto Rico. Thousands of residents were displaced in the fourth largest city in the United States, and although the current situation is beyond difficult, the impacts will be felt for years to come.

And while politicians and climate change deniers will debate whether this hurricane is a result of climate change, for millions around the globe climate change isn't a myth but a reality that threatens their future and their children's future they need to deal with every day. It's clear that more extreme weather events, whether from too much rain or too little, are affecting the economy, farming, conflict, and migration.

But if climate change is the most important global challenge of our time and a threat to our future, it is a particular threat to our food systems. Fortunately, farmers—both small and large—are coming up with their own innovative strategies to improve crop yields, raise incomes, and protect the environment. They're forming cooperatives and growing indigenous crops; they're using cell phone and Global Positioning System (GPS) technologies to get better information about markets, weather, and applying inputs; they're creating equality by valuing women's contributions as food producers and businesspeople; they're mentoring youth and passing on traditional and new skills to the next generation; and they're breaking down silos by working with scientists, nutritionists, researchers, funders and donors, and development agencies to create more participatory research practices.

This book identifies the ingredients that, when combined, can help us to decrease hunger, prevent micronutrient deficiencies, protect water supplies, preserve seeds, prevent food loss and waste, and protect biodiversity. And it highlights the need to invest more in farmers, women, and youth so that they can make the necessary discoveries and innovations.

It is my honor to work closely with the Barilla Center for Food & Nutrition Foundation. As a global research and funding organization, the BCFN is doing important work to develop research on hunger and obesity, nutrition, food loss and food waste, agroecology, global migration, and other projects. And it is building partnerships and collaborations with organizations across the world to help make that research become a reality in fields, kitchens, and boardrooms across the globe.

We all know that preaching to the choir is not the best recipe for change. By bringing this information to a wide audience—of farmers, eaters, businesses, policymakers, academics, youth, funders, media, and civil society—we hope to create the dialogue and the uncomfortable conversations that will help make the food system better for us all.

Danielle Nierenberg is the president of Food Tank and a member of the Barilla Center for Food & Nutrition Advisory Group.

Acknowledgments

We thank Barilla, the Barilla Center for Food & Nutrition, Blue Apron, CARE International, The Christensen Fund, Clif Bar, Clover, Del Mar Global Trust, the Eula Mae and John Baugh Foundation, Fairfield County's Community Foundation, Fair Trade USA, Fazenda da Toca, The Fink Family Foundation, the Food and Agriculture Organization of the United Nations, GRACE Communications Foundation, Great Performances, the International Fund for Agricultural Development, The McKnight Foundation, Naked Juice, Nature's Path, Niman Ranch, Panera Bread, Organic Valley, The Overbrook Foundation, The Republic of Tea, The Rockefeller Foundation, Sealed Air, Sombra Mezcal, the Stuart Foundation, and WhatsGood.

I'm grateful to the Food Tank Board and our Advisory Group for providing much-needed criticism and encouragement for *Nourished Planet*, especially board members Bernard Pollack, our chair, and Dr. William Burke, the newest member of the Food Tank family. Both are my biggest supporters and champions.

I'd also like to thank my parents, Joyce and Fred Nierenberg. They bought me my first notebook and pen and the kid-sized typewriter I tapped out my first stories on. They made me believe I could be a writer and a storyteller, and although my mother continues to be surprised by my dedication to the world's farmers—I couldn't wait to leave Defiance,

Missouri, the farming community I grew up in—she always told me I could be anything I wanted.

And last but not least, this book is for the farmers, small and large, around the globe who work every day to fill our plates and bowls while stewarding the world's biodiversity and other natural resources. I'm grateful that they help nourish both people and the planet.

CHAPTER 1

Food for All

A Recipe for Sustainable Food Systems

IN CHIPATA, ZAMBIA, A REVOLUTION IS TAKING PLACE. The organization Zasaka is getting farmers in that southern African country access to corn grinders, nut shellers, solar lights, and water pumps. Although these technologies might not seem revolutionary, they are producing game-changing results, helping Zambian farmers increase their incomes, prevent food loss and waste, and reduce their load of backbreaking manual labor. But Zasaka is doing more than helping farmers become more prosperous; it is showing the country's young people how farming can be an opportunity, something they want to do, not something they feel forced to do simply because they have no other options.[1]

This project in Zambia is but one ingredient in a recipe for something truly revolutionary: a radically different worldwide food and agriculture system, one built on practical, innovative, and, most important, sustainable solutions to the problems plaguing our current agri-food system.

Farmers, eaters, businesses, funders, policymakers, and scientists are continually learning better ways to increase food's nutritional value and nutrient density, protect natural resources, improve social equality, and create better markets—in short, to develop a recipe for sustainable agriculture for both today and tomorrow. This recipe is being developed in fields and kitchens, in boardrooms and laboratories, by farmers, researchers, government leaders, nongovernment organizations (NGOs), journalists, and other stakeholders in sub-Saharan Africa, Asia, and Latin America. Experts from a variety of socioeconomic and cultural backgrounds are finding ways, firsthand, to overcome hunger and poverty and other problems—while also protecting the environment—in their countries.

Ironically, their recommendations are not that different from those that could be reasonably offered to farmers in North America. Despite all the differences between the developed and developing worlds, there is a growing realization that the Global North's way of feeding people—relying heavily on the mechanized, chemical-intensive, mass production of food—isn't working, and that policymakers and donors might be wise to start following the lead of farmers in the Global South rather than insisting that they follow ours.

In Ethiopia, for example, farmers who are part of a network created by Prolinnova, an international NGO that promotes local innovation in ecologically oriented agriculture and natural resource management, are using low-cost rainwater harvesting and erosion control projects to battle drought and poverty, increasing both crop yields and incomes. In India, women entrepreneurs working with the Self-Employed Women's Association are providing low-cost, high-quality food to the urban poor. In Gambia, fisher folk are finding ways to simultaneously protect marine resources and maintain fish harvests. And hardworking, innovative farmers from all over the world are encouraging more investment in smallholder and medium-holder agriculture and telling policymakers that farmers deserve to be recognized for the ecosystem services they provide, which benefit us all.[2]

There are countless others whose work is showing the world what a sustainable, global food system, or recipe, could look like. They know that the way the food system works today isn't the way it has to work in the future.

They understand that we can help build a food system that combats poverty, obesity, food waste, and hunger, not by treating a healthy environment as an obstacle to sustainable growth but by understanding that it's a precondition for that growth. A food system where science is our servant—not our master—and where it's understood that costly, complicated technology often isn't the most appropriate technology. A food system that honors our values—where women, workers, and eaters all have a seat at the table and none are left on the outside looking in.

The world has a real opportunity and an obligation to build that kind of system, and we don't have a minute to waste. We need to gather the ingredients today so that future generations can build on the recipe for a food system that provides healthful food for all, promotes a healthy planet, and preserves and appreciates food culture.

Ingredients for Sustainability

There are a variety of ingredients to consider if we are to come up with a successful recipe for sustainable agriculture. First, though, we must understand what we mean by "sustainable," which is a term that can encompass so much but is often overused and misused. For us, sustainable agricultural systems are able to efficiently and comprehensively meet the food, fuel, and fiber needs of today without compromising the ability of future generations to meet the needs of tomorrow. What we hope to highlight in this book is how we can create a food system that is not only environmentally sustainable but also economically, socially, and culturally sustainable and that helps ensure that we are nourishing people as well as the planet.

This chapter will introduce the ingredients necessary for sustainability in food and agriculture and explore how the components of sustainability can add up to better food production, more opportunities for farmers, and a healthier planet and population. This section will also discuss in depth a critical ingredient of a sustainable food system: safe and stable access to nutritious food for all. The following sections are

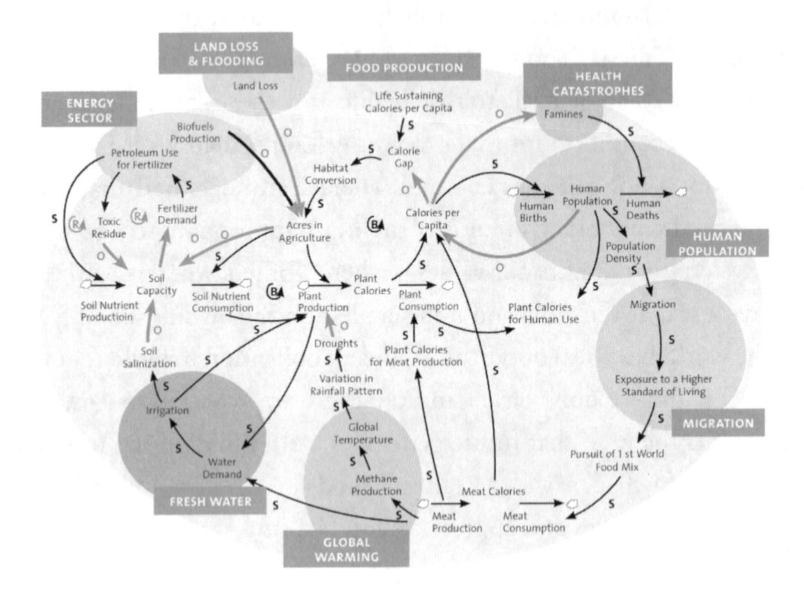

Figure 1.1. Model developed by the International Assessment of Agricultural Knowledge, Science and Technology for Development (IAASTD) to represent the complex system of agriculture. B, balancing; O, opposite; R, reinforcing; S, same. Source: IAASTD, "Agriculture at a Crossroads: Global Report," International Assessment of Agricultural Science and Technology for Development (IAASTD) Project (2008), http://www.weltagrarbericht.de/reports/Global_Report/Global_content.html

devoted to a deeper analysis of the elements crucial to sustainability. This book is much like a cookbook, offering not only a recipe but also insight from experts as well as illuminating real-life examples. Together, this information can provide a well-rounded look at how more sustainable food systems are made.

Exploring sustainability means exploring the foundation on which food production is built (see Chapter 2, "Food for Sustainable Growth"). Air, water, and soil are all important components, and soil—literally *the* foundation of a healthy food system—is the one most often overlooked. Besides being the physical land beneath our feet, soil stores and filters water, provides resilience to drought, and sequesters carbon. One of the biggest threats to the food supply is the loss of topsoil. Indeed, in the past 150 years, roughly half of Earth's topsoil has been lost.[3]

Nearly 40 percent of all land on Earth is used for activities related to agriculture and livestock, and all told, some 4.4 billion hectares (roughly 10.8 billion acres, or about 146 times the area of Italy) is suitable for farming. Yet in the past 40 years, 30 percent of the planet's arable land has become unproductive. In many regions, problems related to soil quality affect more than half of the acreage being cultivated, as seen in sub-Saharan Africa, South America, Southeast Asia, and northern Europe. Each year, the planet loses an agricultural area as big as the Philippines (put another way, we are losing a Berlin-size plot of land every day).[4]

Soil Degradation around the Globe

North America (*Iowa, United States*)

- In Iowa, soil is eroding 10 times faster than regeneration rates.[5]
- Farm fields' ability to retain water has decreased due to sediment loss; the average 5-ton loss of topsoil per acre per year translates to 300–400 gallons of water retention lost per acre.[6]
- Iowa reservoirs are being filled with sediment that has run off from agricultural fields, which reduces their capacity to hold water.[7]
- Soil degradation costs the United States $37.6 billion each year.[8]

Europe (*Italy*)

- Across Europe, soil is eroding 3 to 40 times faster than regeneration rates.[9]
- About 30% of Italy's agricultural area is at risk for erosion, evidenced by an increase in suspended sediment in rivers from runoff and gully erosion.[10]

- Additional soil degradation is caused by a reduction in organic matter, overgrazing, and nonsustainable agricultural practices.[11]

Asia (*China, India*)

- In China, soil is eroding 30 to 40 times faster than regeneration rates.[12]
- In India, economic losses from soil degradation are estimated at 1–7% of agricultural gross domestic product (GDP).[13]

Africa (*Zimbabwe, Niger, Ghana, Kenya, Ethiopia*)

- In Zimbabwe, 70% of farmland has been degraded because of erosion.[14]
- Degraded soils are less responsive to fertilizers because of deficiencies in calcium, zinc, nitrogen, and phosphorus.[15]
- In Niger, 40–50% of land was deforested between 1958 and 1997; current estimates of degradation are 80,000 to 120,000 hectares annually, causing extensive soil loss.[16]
- In Ghana, between 2006 and 2015, soil degradation was projected to reduce agricultural income by approximately US$4.2 billion, or 5% of agricultural GDP over that 10-year period.[17]
- In Kenya, 65% of soil has been degraded, and 19% of agricultural land is experiencing serious degradation.[18]
- The average cost of replacing nutrients to degraded soil is equivalent to 32% of average net farm income in Kenya.[19]
- In Ethiopia, the gross discounted cumulative soil loss (assumes continued degradation over time) ranges from 36% to 44%.[20]
- The rocky northern highlands of Ethiopia receive unpredictable rain, causing extreme erosion that strips nutrients from the soil, devastating crops and livelihoods.[21]

Central America (*Costa Rica*)

- In Costa Rica, uncontrolled soil erosion has led to predictions that coffee yields would decline by half in 3 years and to zero in 20 years, potato yields would decline by 40% after 50 years, and cocoyam yields would decline by more than half after 1 year and to zero after 4 years.[22]

- Soil degradation has reduced agricultural productivity in Central America by 37%—the largest loss of any global region.[23]

South America (*Chile*)

- In Chile, soil degradation has caused a 50% reduction in wheat yields and a 23% decrease in goat rearing.[24]
- Soil degradation and erosion can negatively impact tourism, a major source of income.[25]

Australia (*Australia*)

- Since 1950, populations on "fragile lands" in Australia have doubled; these environments are prone to land degradation and are less suitable for agriculture.[26]
- The livelihoods of agricultural families living on these degraded lands are seriously threatened, placing them in a "poverty-environment trap."[27]

Wes Jackson, an agronomist and the co-founder and president emeritus of the Land Institute, an organization studying the best ways to improve soils worldwide, says we're plowing through our soil bank account and sending those riches downstream to the ocean.[28] Still, there are solutions. For example, the Land Institute touts the value of growing more perennial crops. Agronomist Jerry Glover, formerly at the Land Institute and now part of the Africa RISING project at the U.S. Agency for International Development, also calls for more research into perennial crops, which, unlike annuals, survive from season to season and have deep root structures that can stabilize soils and hold water.[29] Some perennials are also very nutritious, providing an extra source of high-quality food to families in the developing world. According to Glover, more than half the world's population depends on marginal landscapes unsuited for producing annual crops. But perennial crops can be sustainably produced on those lands and improve farmers' yields, he says.[30]

Dr. Sieglinde Snapp, a Michigan State University agronomist who has also worked with the Africa RISING project, strives to create options for smallholder farmers in southeast Africa. Snapp, who focuses on describing what a "greener revolution" could look like, has criticized recent attempts by governments to both suppress perennial crops and "annualize" them in what she sees as an effort to make them more suitable to conventional, chemical-dependent agricultural practices.[31] She writes, "Management approaches that harness biological processes to improve nutrient efficiency are urgently needed. . . . [A]lternative systems can be economically feasible and support sustainable use of resources."[32]

Farmers can also revitalize and protect soils by planting cover crops such as winter wheat, rye, and clover. Besides preventing erosion, cover crops, when plowed under, can increase soil permeability and provide a significant source of soil nutrients for future crops.[33]

In addition, growing diverse crops rather than relying on a single crop, such as corn or soybeans, can restore soil nutrients and help farmers, both large and small, build healthier, more productive soils. Snapp's work has shown that "in Africa, crop diversification can be effective at a countrywide scale."[34] For example, small farmers in the Global South who are raising cattle can use manure to fertilize crops and promote earthworm production. This not only restores nutrients to the soil and protects its microbiota—the microscopic animals that live in soil by the millions—it also helps farmers save money by eliminating the need to buy fertilizer out of a bag and helps mitigate climate change.[35]

Prolinnova works with farmers in Cambodia to fertilize their soil with locally available organic matter. In Nepal, it has installed wasp trappers to encourage beekeeping enterprises, and it is assisting Nepalese farmers as they shift to the more sustainable method of using an ox plow. In

Ghana, Prolinnova has developed salt licks from local minerals for cattle, and in Ethiopia it has engineered a low-cost underground drainage system for frequently waterlogged fields. Prolinnova has also introduced native termite predators as a way of controlling infestations in Uganda. These methods of improving agricultural productivity and soil fertility are inexpensive and sustainable and can be maintained at the local level, providing long-term solutions for farming communities plagued by soil infertility.[36]

"Foodprints," or the ecological footprints or impacts made on the environment—on air, water, and soil—throughout a food's production and distribution processes, are tools crucial to understanding and building sustainability. Dr. John Barrett, of the Stockholm Environmental Institute in New York, attributes the growing awareness of ecological footprints to "the increasing acknowledgment of the environmental impact being placed on other countries by the developed world through their consumption patterns. . . . The ecological footprint provides an overview of the developed countries' dependency on energy and materials."[37]

Box 1.1. Farming the Cities

By 2050, 70 percent of the world's people are expected to live in urban areas, and if we're going to feed all those people, we'll need to continue to make cities and towns into centers of food production as well as consumption. Worldwide, there are nearly a billion urban farmers, and many are having the greatest impact in communities where hunger and poverty are most acute. For example, the Kibera Slum in Nairobi, Kenya, is believed to be the largest slum in sub-Saharan Africa, with somewhere between 700,000 and a million people. In Kibera, urban farmers have developed what they call vertical gardens, growing vegetables, such as kale or spinach, in tall empty rice and maize sacks, growing different crops on different levels of the bags. At harvest time they sell part of their produce to their neighbors and keep the rest for themselves.

The value of these sacks shouldn't be underestimated. During the riots that occurred in Nairobi in 2007 and 2008, when the normal flow of food into Kibera was interrupted, these urban "sack" farmers were credited with helping to keep thousands of women, men, and children from starving.

The role urban farmers played in saving lives in Kibera is probably only a precursor of things to come. In large parts of the less developed world, as much as 80 percent of a family's income can be spent on food. In countries where wars and instability can disrupt the food system and where the cost of food can skyrocket overnight, urban agriculture can play a fundamental role in helping prevent food riots and large-scale hunger. In that respect, promoting urban agriculture isn't only morally right or environmentally smart, it's necessary for regional stability.

But urban agriculture isn't important only in sub-Saharan Africa or other parts of the developing world. In the United States, AeroFarms runs the world's largest indoor vertical farm in Newark, New Jersey, where it grows greens and herbs without sunlight, soil, or pesticides for local communities in the New York area that have limited access to greens and herbs. Another group, the Bronx Green Machine, which is based in New York City's South Bronx neighborhood, is an after-school program that aims to build healthy, equitable, and resilient communities by engaging students in hands-on garden education.[a]

Across the Atlantic, in Berlin, Germany, a group called Nomadic Green grows produce in burlap sacks and other portable, reusable containers. These containers can be set up in unused space anywhere, ready to move should the space be sold, rented, or become otherwise unavailable. In Tel Aviv, Israel, Green in the City is collaborating on a project with LivinGreen, a hydroponics and aquaponics company, and the Dizengoff Center, the first shopping mall built in Israel. This collaboration provides urban farmers with space on the top of the Dizengoff Center to grow vegetables in water, without pesticides or even soil. Green in the City also provides urban farming workshops and training in the use of individual hydroponic systems.[b]

---

a. "Our Farms," AeroFarms (2017), www.aerofarms.com/farms/; "The Green Bronx Machine Grows Organic Citizens," Food Tank (2016), https://foodtank.com/news/2016/06/the-green-bronx-machine-grows-organic-citizens/

b. "Twelve Organizations Promoting Urban Agriculture around the World," Food Tank (2016), https://foodtank.com/news/2016/12/twelve-organizations-promoting-urban-agriculture-around-world/

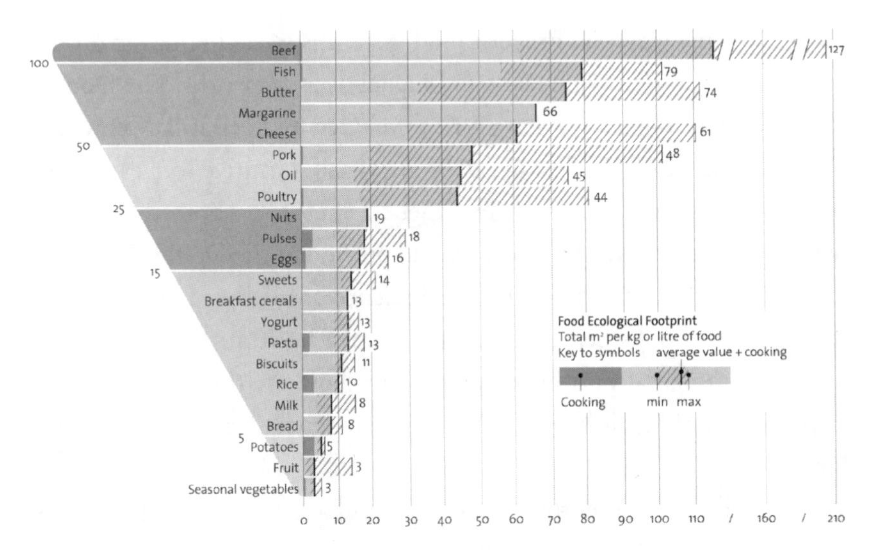

Figure 1.2. The ecological footprint of foods (global square meters per kilogram or liter of food). Source: BCFN Foundation, 2015, from p. 104 of *Eating Planet,* 2nd ed. (Milan: Edizioni Ambiente and the Barilla Center for Food & Nutrition, 2016).

As Earth's supply of freshwater dwindles, its temperature is on the rise. Climate change's effects are being seen everywhere, from the relentless melting of the Greenland and Antarctic ice sheets to the increasing acidification of the ocean's surface waters, which are 30 percent more acidic today than at the start of the Industrial Revolution.[38] According to the Organisation for Economic Co-operation and Development, agriculture is a top contributor to the climate crisis, because more than 14 percent of global greenhouse gas (GHG) emissions come directly from the growing crops. According to the CGIAR Research Program on Climate Change, Agriculture, and Food Security, that number climbs to between 19 and 29 percent of global emissions when the rest of the food system's impacts are considered, including land use change for agricultural purposes, the production and use of synthetic fertilizers, and the transportation of inputs and outputs.[39]

At the same time the modern food system is wreaking havoc on the environment, obesity rates are climbing all over the world. The Double Pyramid (see Chapter 2, Figure 2.1), developed by the Barilla Center for Food & Nutrition (BCFN) in 2010, is a visual, intuitive tool eaters can use to calculate the effects of their dietary choices on both the planet's health and their own. The food pyramid portion of the Double Pyramid, which is based on current nutritional science, visually represents which food groups should be consumed most for a healthful, balanced diet. The environmental portion is designed to showcase the environmental impacts, including GHG emissions and water consumption, of different food groups.[40]

The BCFN describes the Double Pyramid as a visual tool that allows eaters to see that "the foods that are advised to be eaten more are also, generally, those that have the lowest environmental impacts. On the other hand, foods that are advised to be eaten less are also those that have a greater environmental impact."[41]

A recent study published in the *Proceedings of the National Academy of Science* reveals that in the United States, beef production requires 28 times more land than pork or chicken production and 11 times more water and creates five times more GHG emissions. Along with its negative impacts on the environment, the fat and cholesterol in beef can be bad for the eater's health.[42]

The final ingredient is the observation of food culture: honoring the traditions and indigenous knowledge around agriculture and recognizing the nutritional importance of the Mediterranean diet (See Chapter 4, "Food for Culture"). Food can be as important to a culture as language. Not only is it a tool for communication, it is also a source of social engagement, the foundation of many religious practices, and a platform for human expression. Understanding the culture of food also means acknowledging the important role of women and youth in creating a more sustainable food system.

Box 1.2. The Regenerative Agriculture Movement

Regenerative agriculture involves farming practices that can help mitigate climate change and the adverse impacts of industrial agriculture by rebuilding and restoring degraded soil and improving biodiversity. Organizations and farmers around the world are currently using regenerative methods on an ever-larger scale and are encouraging others to follow suit by educating producers and other stakeholders about the benefits of regenerative agriculture.

Kiss the Ground, a nonprofit whose goal is to regenerate soil on a global scale, is teaching people around the world how to turn biodegradable waste into healthy soil. The group believes compost can play a crucial role in restoring balance to the world's carbon cycle by regenerating soil that will sequester excess carbon from the atmosphere.

In 2017, a like-minded organization, Regeneration International, launched the Regeneration Hub in collaboration with Open Team and 17 other regenerative farming organizations. This hub hopes to educate farmers and eaters about the benefits of the regeneration movement and inspire collaboration between farmers and environmental organizations.

Similarly, the nonprofit Rodale Institute launched a worldwide campaign to spread awareness and education about regenerated soil's ability to reverse the effects of climate change, and it is calling for the restructuring of the food system on a global scale. The institute's white paper on the campaign outlines its goals and states, "We could sequester more than 100 percent of current annual CO_2 emissions with a switch to widely available and inexpensive organic management practices, which we term 'regenerative organic agriculture.'"[a]

a. J. DeMarco, "Kiss the Ground Releases the Compost Story with a Celebrity Cast," Food Tank (2017), https://foodtank.com/news/2017/05/the-compost-story-brought-to-you-by-kiss-the-ground/; A. Groome and J. Wilson, "The Regeneration Hub: Mapping the Regeneration Movement," Food Tank (2017), https://foodtank.com/news/2017/03/regeneration-hub-mapping-regeneration-movement/; "Regenerative Organic Agriculture and Global Change: A Down-to-Earth Solution to Global Warming," The Rodale Institute (2014), https://rodaleinstitute.org/assets/RegenOrgAgricultureAndClimateChange_20140418.pdf

Food for All

Access to food is one of the first and most fundamental of all human rights, and it is essential to the well-being of any society. Unfortunately, adequate food is more accessible to some people than to others. The final ingredient in our recipe for a sustainable food system is correcting the imbalance of food access found in populations around the world.

The most commonly accepted definition of food security identifies it as a condition in which all people, at all times, have physical and economic access to sufficient, safe, and nutritious food. The concept of food security thus includes four components: the availability of sufficient quantities of quality food, adequate means of access to such food, the maintenance of sanitary conditions to allow the safe consumption of food, and the stable, consistent availability of food over time. Unfortunately, too many people cannot claim even one of these components of food security in their lives, even as other populations seem to take them for granted. It is the urgency of correcting this imbalance that drives the need for farmers and policymakers to embrace the cause of more sustainable food systems.[43]

Poor production and consumption practices and the insufficient distribution of food can lead to hunger and even famine. In fact, even as this book was being written, at least four famine situations—in Yemen, South Sudan, Nigeria, and Somalia—threatened at least 20 million people.[44]

Undernourishment and malnutrition, suffered mostly by the poor in developing countries that often lack basic healthcare and clean water, can have devastating economic and social impact. Malnourished citizens are often too ill to work, and the issues of access to and control of agricultural resources are a source of political and social tensions. Furthermore, as climate change aggravates the viability and security of agriculture and food production, food and water issues can exacerbate unresolved ethnic, religious, and economic tensions. The costs of undernourishment come primarily from lost productivity and a reduced contribution to the workforce due to limited schooling and early death. In Ecuador and Mexico, for example, the economic impact of malnutrition has risen to 4.3 percent and 2.3 percent, respectively, of the GDP, which creates a financial burden one and a half to three times greater than the economic burden placed on the United States by the obesity epidemic.[45]

It is important to emphasize that access to food is not a problem exclusive to developing nations; it is an issue in the industrialized world as well. According to a 2012 report by the UN Food and Agriculture Organization (FAO), the number of undernourished people in developed countries reached 16 million in 2012, an increase of 3 million from 2006. Interestingly, although studies originally predicted that the 2008 recession in the United States and other developed countries would cause a sharp increase in hunger in the developing world, this increase was far less pronounced than expected, as some governments were largely successful in efforts to soften the effects of the crisis.[46]

Each of the four components of food security can be addressed at the individual, community, local, national, and global levels. At the individual level, farmers can practice sustainable agricultural methods, replacing the unsustainable methods that rely on fossil fuels and high levels of inputs, such as the use of harmful fertilizers and inefficient, large-scale irrigation, and that ultimately only increase food insecurity. At the community and local levels, greater crop diversity can lessen food insecurity, as the unsustainable practice of monocropping increases the chances of disease and famine. At the national and global levels, policy changes can help reduce food insecurity by promoting incentives for farmers to grow sustainably, and the expanded use of food stamps at farmers' markets can allow all people the opportunity to buy fresh, local, and healthful food.[47]

Although the world produces enough food to feed its entire population, the imbalance in food access has continued to expand over the past decade, creating a revealing paradox: The world is both underfed and overfed. There are 2.1 billion obese or overweight people in the world, in

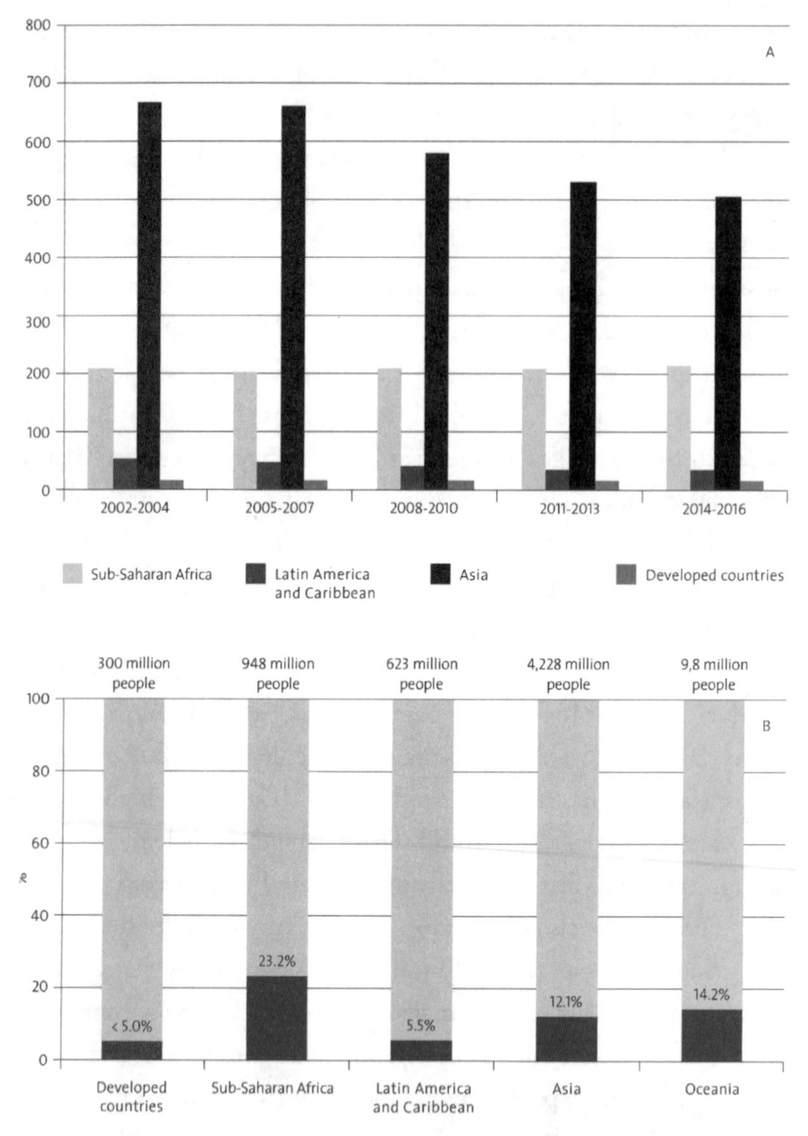

Figure 1.3. Worldwide undernourishment situation (2014–2016 values are estimated). A: Undernourished people in selected regions of the world. B: Percentage of undernourished people of the total population of each region of the world (2014–2016). Source: FAOSTAT Food Security Indicators, 2015, in FAOSTAT Database, Food and Agriculture Organization of the United Nations (2015), http://faostat3.fao.org/home/E

developing and industrialized countries alike. At the same time, at least 815 million people around the globe are hungry. In the United States, more than 30 percent of the population is considered either obese or overweight, and obesity now affects 70 percent of Mexico's population. Even in India, the poster child for world hunger, 17 percent of adults are obese. The growth in obesity has been accompanied by a surge in cardiovascular and respiratory diseases and type 2 diabetes. Besides the obvious health consequences of this situation, there are its dire financial effects: Treating disease will cost US$30 trillion globally between now and 2030. Ironically, those who are obese are often also malnourished because their diet is based on starchy, processed foods that can be high in fat and sodium and low in nutritional content.[48]

Food insecurity as a cause of obesity and poor public health can be explained in several ways. First, food-insecure people often can afford only less expensive, calorie-dense foods, which tend to cause weight gain. Second, people often overeat after periods without enough food. Third, fluctuating eating habits can confuse the body's metabolic system and cause weight gain even when people aren't eating more calories.[49]

The coexistence of vast numbers of malnourished people and the quantity of food produced worldwide creates a revealing paradox, disproving the popular belief that hunger can be resolved by simply producing more food. Revolutionizing the food system as we know it is the solution. Kostas G. Stamoulis, a senior economist at the UN's FAO, states, "The way we manage the global agriculture and food security system doesn't work. There is this paradox of increasing global food production, even in developing countries, yet there is hunger."[50] By examining the structural issues of the imbalances in food access, advocates can create better solutions.

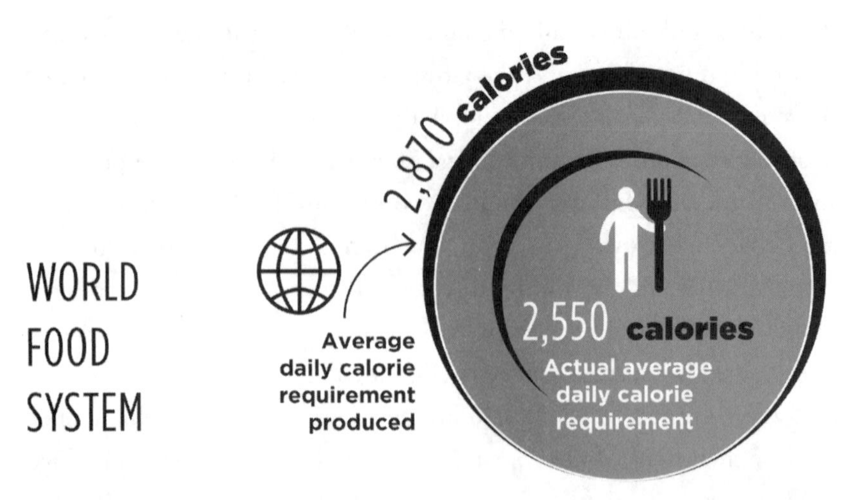

The world food system is capable of producing today a little less than 2,870 calories per person per day, compared with an actual average per capita calorie requirement for an adult individual of 2,550 calories

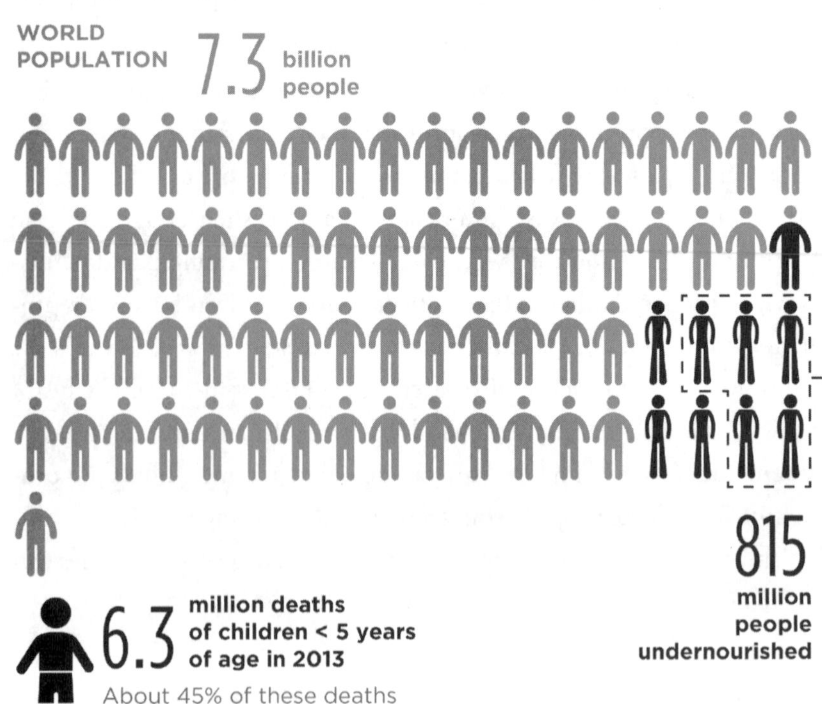

Figure 1.4. Hunger versus obesity.

HUNGER VS. OBESITY

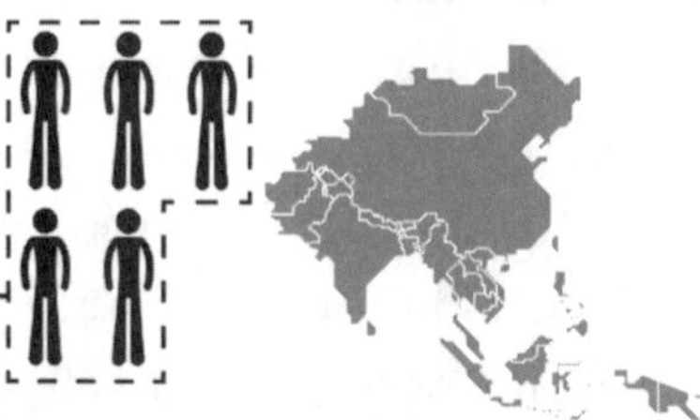

490 **million hungry people live in Asia**

23% **of the sub-Saharan population is undernourished**

62% of the world's hungry live in Asia, a region where there is also an increase in the number of overweight or obese people. In ub-Saharan Africa, the percentage of undernourished people in the total population is the highest in the world (23%). Although there has been a decline in the percentage of undernourished people compared to 1990-1992, the absolute number of people continues to increase due to population growth

2.1 **billion obese or overweight in 2015**

In 2013, 45 million children < 5 years of age were obese or overweight

33 **million deaths annually from diseases linked to excess food**

1/3 **FOOD WASTE**

One third of global food production is lost, destroyed or wasted during preservation, processing, distribution and consumption

Barriers to Food Access and Affordability

Several structural issues underlie the global imbalances in food access, but the primary one is poverty. As of 2013, 10.7 percent of the world's population was living on less than US$1.90 a day. In addition, the world's population is becoming more urban, and by 2016, 54.5 percent of people, or more than 4 billion, were living in urban and periurban areas. This means that the rest of the population, or roughly 3.4 billion people, are living in rural areas. Many of them are poor, and most of them, ironically, are farmers.[51]

Although some advocates for the poor call for more equal distribution of wealth or the establishment of higher minimum wages or living wages in the fight against hunger and poverty, economic growth through farming and agriculture represents a real and sustainable solution. The critical targets in the battle against inadequate nutrition—an increase in wealth and a more equitable distribution of that wealth—are among the best but least known economic tools for addressing these problems. According to the "World Development Report" published in 2008 by the World Bank, any percentage-point increase in the GDP generated by the agricultural sector is twice as effective in terms of reducing poverty as equivalent growth in other sectors.[52]

Farming is important in the fight against poverty for two big reasons. First is the high incidence of poverty in developing countries, where agriculture is the primary source of income and employment. Second is the fact that rural populations typically possess lower levels of education and less access to alternative sources of income (such as employment in manufacturing or services). This means that the agricultural sector makes up a key element in the development of strategies aimed at improving the well-being and livelihoods of those living in rural areas.

But it isn't enough for the Global South to invest solely in increasing agricultural productivity; instead, what is needed, according to an International Monetary Fund report titled "Causes and Consequences of Income Inequality: A Global Perspective," is a more equal distribution of wealth through the creation of job opportunities for the poorest segments of the population. Without a wider distribution of wealth, farming in many poor and developing nations is destined to remain purely at subsistence level.[53]

Each year, an estimated 1.3 billion tons of food is wasted or lost across the world. In developing countries, that can be as much as 30 to 40 percent of all food produced. So investments in infrastructure, including better roads, cooling facilities for produce and dairy products, and better storage for grains and other crops, are essential (see the section on food loss and food waste in Chapter 2).

Among the other structural factors that significantly affect access to food is the growing competition for agricultural land and water resources, which is amply demonstrated by the acquisition of large tracts of land by foreign investors, often in partnership with local and national governments. These lands, which many times have been used for generations by pastoralists and farmers, are converted from traditional uses by the local community to commercial uses (in a practice known as land grabbing).

From a policy standpoint, national subsidies for agriculture often encourage the production of commodity crops on large farms while ignoring the needs of small and medium-size farms that grow fruits, vegetables, and more nutritious grains. Raising crops for biofuels and maize and soy for animal feed have also meant that huge areas are being devoted to fueling cars and feeding livestock rather than nourishing people. And there continues to be a lack of political will among both international and national leaders to address poverty and hunger in a systemic way. Agriculture and food security are often relegated to a position of secondary importance, and resources are channeled into food aid rather than into programs that will actually benefit family farmers across the globe.

Other factors besides structural ones have thwarted progress toward a poverty- and hunger-free world. The difficulty of forecasting the future of markets for agricultural products and foodstuffs is partly the result of a complex set of phenomena on a global scale, including worldwide demand and new opportunities in the agro-food market (e.g., energy, green chemistry), the effects of climate change, economic and demographic growth, uncertain environmental conditions, and the financialization of agricultural commodities.[54]

With respect to climate change, for many experts, including those at the FAO, the answer is a two-pronged strategy: mitigation and adaptation. The more we combat climate change through a broad, shared consensus, the more that agricultural strategies will be able to address mitigation, primarily through adaptation. Examples of adaptive techniques that can be used to address mitigation include planting cover crops in between other crops such as maize or soybeans, which improves nutrient cycles and reduces soil compaction and erosion; and agroforestry, or growing trees along with crops, which can help farmers conserve water, prevent soil degradation, and add nutrients to the soil.

Agriculture can be an essential part of the solution to climate change through approaches that seek to grow agricultural output in a sustainable manner and increase resilience to environmental pressures, reducing GHG emissions and encouraging farmers to adopt new practices to adapt to this new reality.[55]

In addition, the decline in public and private investment in agriculture over the past 30 or 40 years and the corresponding lack of political interest has created the illusion that our food system can be managed without funding from national governments and the funding and donor communities. Unfortunately, it took the 2007 and 2008 food and financial crisis and the resulting food riots and other political instability for donors and governments to take notice. In fact, the growth in food exports after the recession was roughly twice the average between 1998 and 2007, and developing countries' share of U.S. agricultural exports had increased to more than 60 percent by 2011.[56]

Box 1.3. Grabbing Land

The Oakland Institute, an independent policy think tank specializing in social and environmental issues, defines land grabs as purchases of large tracts of land in poor, developing countries made by wealthier nations and private investors. The occurrence of land grabs, which is rising rapidly, is caused by a combination of global food price volatility, global food crises, and growth in speculative activity. More factors are fueling the spread of land grabs as well, such as the rise in land investments, the demand for agrofuels, and the rush of food-insecure countries to accumulate food banks.[a]

A recent Oakland Institute report examines the impacts on global food security as international financial institutions drive control of food sources away from the public sector and to the private sector, severely compromising poor nations as they try to achieve food self-sufficiency.[b] Among these institutions, the International Financial Corporation (IFC), the private-sector arm of the World Bank Group, plays a major role, financing private investments in the developing world by advising governments and businesses and encouraging governments to facilitate business transactions.[c]

In 2008, the government of investment-starved Pakistan put 1.1 million acres of agriculture land up for sale and, in line with IFC philosophy, changed its laws to make it easier and more financially appealing for foreign investors to buy the land. Firms investing in Pakistani land markets are now offered a range of incentives, from 10-year tax breaks to assurances that high profits will be legally protected.[d]

a. S. Daniel and A. Mittal, "The Great Land Grab: Rush for World's Farmland Threatens Food Security for the Poor," The Oakland Institute (2009), https://www.oaklandinstitute.org/sites/oaklandinstitute.org/files/great-land-grab.pdf

b. Ibid.

c. Whereas the World Bank (International Bank for Reconstruction and Development and International Development Association) provides credit and nonlending assistance to governments, the IFC provides loans and equity financing, advice, and technical services to the private sector. The IFC is one of the fastest-growing institutions in the World Bank Group and has made important investments to improve local private sector companies, but its projects have often been carried through at the expense of physical and economic displacement for thousands. Source: "International Financial Institutions," Bank Information Center (BIC) (2017), http://www.bankinformationcenter.org/resources/institutions/ifc/

d. A.F. Khan, "Corporate Farming and Food Security," *DAWN* (29 December 2008), https://www.dawn.com/news/336215

The absence of regulation on corporate investors as they buy up land has created a grave situation for small farmers in Pakistan's Punjab province. A proposed land deal with the United Arab Emirates could displace 25,000 villagers.[e] Of the 6 million families working approximately 50 million acres of land, 94 percent are considered subsistence farmers, each occupying an average of less than 12.5 acres.[f] Furthermore, much of the land they have been farming for generations is owned by large private holdings or the government.[g]

In Cambodia, government incentives and an unusually stable economy are contributing to a wave of foreign investments and fierce competition for fertile farmland, exacerbating the burdens of rural landlessness and food shortages. The Cambodian government continues to clear a path for foreign investors such as Kuwait and the United Arab Emirates, claiming that such land deals can solve the country's food crisis. They support these claims with the statistic that only 2.5 million of Cambodia's 15 million acres of available arable land are being used.[h] Meanwhile, according to estimates, tens of thousands of Cambodians have been displaced in recent years.[i] Today, says the FAO, 33 percent of Cambodian citizens are undernourished in the face of widespread rural poverty.[j]

Some 1.5 billion small farmers in the world live on less that 2 hectares of land.[k] Their food security, as well as that of rural populations throughout the developing world, depends on their having secure, equitable access to and control over the land they use to produce food.

---

e. Ibid.

f. Ibid.

g. I. Husain, "Pakistan, the Great Land Grab," *DAWN* (9 May 2009), https://www.dawn.com/news/833579

h. D. Montero, "Insecurity Drives Farm Purchases Abroad," *Christian Science Monitor* (22 December 2008), https://www.csmonitor.com/World/Asia-South-Central/2008/1222/p01s06-wosc.html

i. Ibid.

j. "The State of Food and Agriculture (SOFA), 2008," Food and Agriculture Organization (FAO) of the United Nations (2008), http://www.fao.org/docrep/011/i0100e/i0100e00.htm

k. "Kathmandu Declaration: Securing Rights to Land for Peace and Food Security," International Land Coalition (2009), http://www.landcoalition.org/sites/default/files/documents/resources/09_katmandu_declaration_e.pdf

Precarious Prices for Food

The instability of food prices can create a barrier that prevents those in need from gaining better access to healthful and nutritious food.

Overall, the highest proportion of the chronically hungry are poor, and many of them are small farmers. Volatile or unstable food prices can put those with low incomes in the poorest countries at the greatest risk, because they spend the highest proportion of their incomes on food of any group in the world.[57] The World Bank estimates that the 2007–2008 world food price crisis kept or pushed 105 million already-poor people below the poverty line in low-income countries.[58]

In 2008, and again in 2010, global food prices soared, plunging millions worldwide into poverty. Riots occurred in at least 48 countries, according to the World Food Programme (WFP), and millions of people joined the ranks of the hungry. In developing countries, where many people spend 60 to 80 percent of their income on food alone, higher food prices led to widespread malnutrition and civil unrest.[59]

The phenomenon of food price volatility—the variation, or rapid jump, in food prices, especially over a short period of time—has taken a dramatic toll on eaters in poor countries over the past decade. From July 2010 to February 2011, the FAO Food Price Index, which measures world food prices, skyrocketed by 38 percent, topping a peak reached during the 2007–2008 food and financial crisis. Although increases in yields over the past few years have stabilized the global food price index, policymakers, farmers, and funders and donors need to remain vigilant in monitoring world food prices and their sudden, unpredictable changes.[60]

To understand how food prices affect developing countries and what must be done at a policy level, it is important to separate the interconnected issues that determine food prices. Stock markets, politics, social stability, and climate change are four of the large issues that together create instability in food prices.[61]

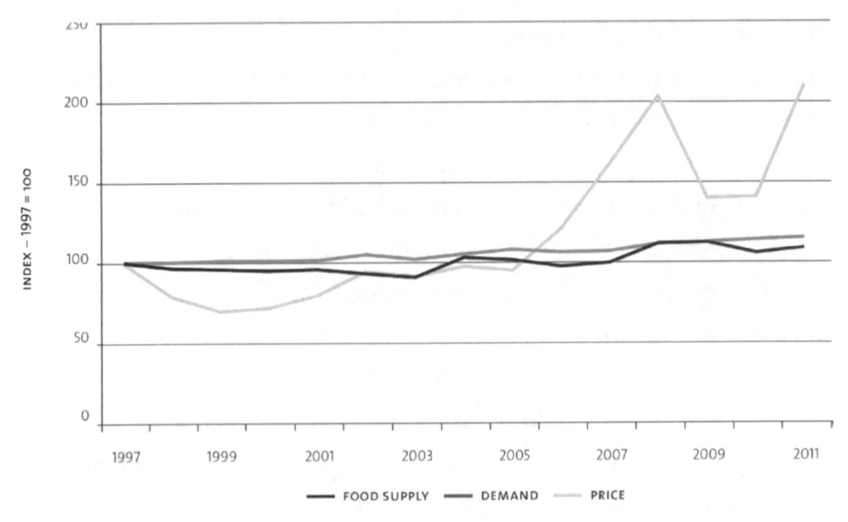

Figure 1.5. Correlation between oil prices and food prices (January 1997–April 2011). Source: BCFN on FAO and IMF data, 2011.

Food markets are distinctive in that both supply and demand curves are unpredictable. Large price changes can occur over time, influenced by actions and incidents economists refer to as "shocks." Shocks are now more frequent, resulting in greater price volatility. Supply shocks can be caused by extreme weather (such as droughts in Russia and China and floods in Australia and South Africa), the production of biofuels, and the scarcity of natural resources. Demand shocks can result from unanticipated policy decisions, social upheavals, and sudden economic growth in developing countries.[62]

The financial markets play a large role in demand shocks. According to recent articles in *Time* and *The Atlantic* and a report from the Institute for Agriculture and Trade Policy, financial speculation has undermined the commodities market's ability to act as a stabilizing force for farmers, buyers, and consumers. Food prices once again reached 2008's record levels in 2012, according to the FAO, and Oxfam has warned that the price of staples such as wheat and rice may double in the next 20 years.

These prices are no longer driven by simple supply and demand but instead by the activities of financial speculators.[63]

In the past, commodity prices were determined by such real-life factors as quality and supply and demand. Now it is largely *perception* of the market that determines the price of a commodity. In 2008, for example, investors' trepidation about the drought in Australia affected the pricing of all wheat, even different varieties grown elsewhere in the world. These fears turned out to be unfounded; in fact, 2008 was an excellent year for wheat. Data from the U.S. Department of Agriculture (USDA) found that 657 million bushels of wheat remained in silos after the buying season.[64]

Financial speculators now dominate the commodities market, holding a hefty 60 percent of some market shares compared with the more modest 12 percent they held 15 years ago. They have plenty of incentives to continue business as usual, as Brett Scott, a writer for *The Ecologist*, points out: "Goldman Sachs [is] estimated to have made more than $1 billion in 2009 alone and Barclays as much as £340 million [about $US441 million] a year from trading food commodities." Commodities now account for more than $126 billion in investments.[65]

Politics and policymakers have a twofold effect on price volatility: They can cause problems, and they can also offer solutions. Sophia Murphy and Timothy Wise, authors of "Resolving the Food Crisis: Assessing Global Policy Reforms Since 2007," found that responsible regulation, such as 2012's Commodity Exchange Act, can inhibit excessive financial speculation and prevent it from driving food prices higher. Still, the success of regulations such as these depends largely on international cooperation and on how successful lobbyists are in their efforts to undermine the legislation by claiming other factors are to blame for the food price fluctuations.[66]

Frederick Kaufman, a professor of journalism at the City University of New York and a writer for *Harper's* magazine, says,

> When it comes to crop insurance and other supports, we have to separate the farmer from agribusiness. One reason agribusiness has not been vertically integrated is that industrialists do not want to bother with the risk of putting seeds in the ground and praying for rain. And while farming may be for gamblers, national security rests on a steady and reliable food supply. Historically, America has understood the need to support farmers. The country has enjoyed years of prosperous harvests of inexpensive wheat because of federal dollars spent on agricultural education and outreach. Washington is right to support those in a high-risk business essential to our national security and international trade. History has shown that when farmers go belly-up, the effects reverberate throughout society. The last thing we want in these days of drought, flood, and climate change is for this country to become dependent on Argentina, Brazil, and Russia for our daily bread.[67]

The effects of social conflicts on the volatility of food prices are also twofold: A shortage of food can cause political and social tension as well as exacerbate it. For example, in 2013 a perfect storm of events occurred in Egypt, offering increasing evidence of the impacts of climate change on agriculture, migration, and high food prices. Although experts from the Center for American Progress, the Center for Climate and Security, and the Stimson Center agree that the Arab Spring was apt to happen no matter what, it probably began earlier because of the confluence of those events.[68]

Extreme weather patterns resulting from climate change are a major source of food supply shocks. In 2010, droughts in both Russia and the United States led to higher prices for maize and soybeans as well as wheat.[69]

According to the Intergovernmental Panel on Climate Change (IPCC), the foremost international authority on the impact of climate change, it is very likely that more than half of global warming is the result of human activity. Rising temperatures, the changing frequency and overall amount of rainfall, and above all the intensification of extreme weather events have an enormous impact on agricultural production. The occurrences limit the ability of farmers to maximize yields and total agricultural production, putting food availability—one of the essential components of food security—at risk. This can also cause food prices to rise in the medium to long term.[70]

The IPCC's Fourth Assessment Report indicated that worldwide wheat yields could drop by 2 percent each decade, because wheat is so sensitive to heat. It also projects that fish catch will fall between 40 and 60 percent in parts of the tropics. The report also expects rising temperatures and changing rain patterns to raise food prices by 3 percent to as much as 84 percent by 2050. (This large range in the projected percentage change probably reflects the variation in long-term climate change projections and variations between crops.) These increases in food prices could create political instability similar to the riots seen in Asia and Africa after the 2008 food price shocks.[71]

Food prices are part of a complex system that doesn't always work, and the system needs major repairs if it is going to become more sustainable and predictable. Strengthening local and regional food systems and investing in long-term solutions to mitigate and adapt to climate change are the most effective ways to curb prices and prevent political instability. Furthermore, the global adoption of best practices is necessary, as is defining a new system of rules for the commodities markets and international trade policies.

Institutions such as the International Fund for Agricultural Development (IFAD) and the FAO are working to help farmers and eaters alike manage the uncertainty of high and volatile food prices. In partnership with the IFAD and the FAO, the WFP released a joint statement at the G20 Summit in Mexico calling on the G20 to redouble their efforts to fight global hunger: "We, the United Nations food and agriculture agencies—FAO, IFAD, and WFP—welcome the priority given to food and nutrition security by the Mexican Presidency of the G20, keeping it at the forefront of the global development agenda." The statement argued for "the continuing recognition by the G20 of the pivotal role of smallholder agriculture to global food security and to boosting productivity in a sustainable manner," and it urged for "food and nutrition security to remain prominent on the G20 agenda in the coming years."[72]

Nearly a billion people worldwide are hungry and at the mercy of volatile food prices, and it is evident that the causes lie within the global food system. Barbara Stocking, Oxfam's chief executive, said, "The food system is pretty well bust. All the signs are that the number of people going hungry is going up. One in seven people on the planet go hungry every day despite the fact that the world is capable of feeding everyone. The food system must be overhauled if we are to overcome the increasingly pressing challenges of climate change, spiraling food prices, and the scarcity of land, water, and energy."[73]

Conclusion and Action Plan

One consequence of the Green Revolution is that over the past 60 years, agricultural production in Africa, Asia, Latin America, and Europe and the United States has become more concentrated on the production of raw commodities, including corn, soy, wheat, and other crops, and less on more nutritious foods, such as millet, sorghum, perennial grains, and vegetables.

Although it's true that the poor in developing nations get most of their calories from starchy crops, there's been very little funding for research into what makes those starchy staples palatable. In fact, funding for research on cereals and grains is now roughly 13 times the amount spent on research on fruits and vegetables. And that lack of investment comes at a high price.

Box 1.4. Creating Deserts

For millions of Americans, something as simple as finding a fresh apple to snack on can be a challenge. In recent years, many parts of this country have seen significant declines in the number of grocery stores or other outlets where fresh fruits, vegetables, grains, and meats are available. At the same time, the number of convenience stores, liquor stores, and fast-food chains selling processed foods high in fat and sugar has increased.

These so-called food deserts are geographic areas where access to healthful, affordable food options (particularly fruits and vegetables) is restricted or nonexistent because of the absence of full-service grocery stores within convenient traveling distance. The USDA defines a convenient traveling distance as no more than 1 mile in urban areas and no more than 10 miles in rural areas. This is not about access to any food; it is about access to and ability to afford healthful foods, especially among vulnerable populations. The USDA estimates that nearly 23.5 million Americans live in food deserts and that 13.5 million of them are low-income residents. Across the country, low-income ZIP codes have 25 percent fewer supermarkets and 30 percent more convenience stores than middle-income ZIP codes.[a]

Food access issues affect urban and rural areas alike. Those who do not have access to a car find it impossible to reach stores beyond their immediate neighborhood and must rely on friends, neighbors, and food pantries for meals.

Solutions must go beyond the idea of simply building more grocery stores, because both rural and urban food desert areas have a hard time attracting and keeping commercial grocery retailers. According to the Center for Rural Affairs, one in five grocery stores in rural areas has gone out of business in the last 4 years. As more people realize there's no simple, uniform solution to the food desert issue, they have come up with varied and innovative strategies for eliminating food deserts. Here are five of them.[b]

- When Brahm Ahmadi was unable to get private investors interested in his plans for a new grocery store, he went directly to the public for funding.

a. "Food Desert Locator," U.S. Department of Agriculture Food and Nutrition Service (USDA FNS) (2011), https://www.fns.usda.gov/tags/food-desert-locator

b. J. Bailey, "Rural Grocery Stores: Importance and Challenges," Center for Rural Affairs (2010), www.files.cfra.org/pdf/rural-grocery-stores.pdf

People's Community Market will provide a full-service neighborhood food store and a health resource center and community hub to 25,000 West Oakland residents who have limited access to fresh produce.[c]

- Also in California, a new public–private partnership loan fund called Fresh Works is drawing on a US$200 million investment pool to provide loans and grants to grocers looking to build or expand in underserved neighborhoods.[d]
- Community garden initiatives are quickly popping up in both urban and rural areas, providing easy access to inexpensive, fresh produce. The American Community Garden Association provides resources to more than 18,000 community gardens in the United States and Canada.[e]
- After a survey revealed that 94 percent of residents would purchase more fresh produce if it were available at convenience stores, the city of Minneapolis enacted the Minneapolis Health Corner Store initiative, which requires all corner and convenience stores to stock a certain amount of fresh fruit on their shelves.[f]
- Mobile markets and produce trucks have appeared in many underserved areas. In California, since 2006, Second Harvest Food Bank has been using a refrigerated truck to distribute fresh produce to those in need. In Detroit, Peaches & Greens, a mobile produce truck, delivers fruits and vegetables to a high-need community. Garden on the Go, a produce truck initiative from Indiana University, makes 16 weekly stops, sells produce at reasonable prices, and accepts food stamps, making access to healthful fruits and vegetables as convenient and affordable as possible.[g]

---

c. "People's Community Market: More than a Grocery Store," People's Community Market (2017), www.peoplescommunitymarket.com

d. "California FreshWorks Fund," Northern California Community Loan Fund (2017), https://www.ncclf.org/california-freshworks/

e. "American Community Gardening Association," American Community Gardening Association (2017), https://communitygarden.org

f. "Healthy Corner Stores," City of Minneapolis Government (2017), www.minneapolismn.gov/health/living/new%20cornerstores

g. "Second Harvest Food Bank of Santa Clara and San Mateo Counties," Second Harvest Food Bank (2014), www.shfb.org; "Peaches and Greens Produce Market," Peaches and Greens Produce Market (2017), www.peachesandgreens.org; "Garden on the Go," Indiana University Health (2017), www.iuhealth.org/about-iu-health/in-the-community/healthy-weight-and-nutrition/garden-on-the-go/

Dyno Keatinge, former director general of the World Vegetable Center in Taiwan, says that the lack of vegetables in children's diets has a catastrophic impact on childhood malnutrition and mortality. Simply put, the more countries have access to vegetables, the fewer children will die before the age of 5. And obviously, if children aren't getting vegetables to eat, they're at much greater risk of malnutrition and stunting. This not only curbs their personal and educational development but also has serious implications for a country's economic development.

Although shifting the food system's emphasis from starchy staples or processed foods to vegetables is essential to promoting nutrition and public health, it's also key to protecting the environment and reducing poverty.

For example, in Uganda, Developing Innovations in School Cultivation, or Project DISC, is helping young people see agriculture as a career. Its founder, Edie Mukiibi, the vice president of Slow Food International, is working with more than two dozen schools to teach kids from preschoolers to teens how to grow and cultivate—and market—indigenous fruits and vegetables. One spinoff of Project DISC's work is that it's restoring a demand for traditional food that had been long forgotten (see Chapter 4).

The final ingredient of a more sustainable food system is cultivating greater equality. It's absurd that although women make up more than half of the world's population and nearly half of its farmers, their contributions to agriculture are at best largely unnoticed and at worst almost universally ignored.

As we make headway, we'll be able to see progress, because as goes the fate of women, so goes the fate of the world. According to the FAO, if women farmers had equal access to resources—land, credit, education, extension services—these workers could increase their food production by 20 to 30 percent and lift as many as 150 million people out of hunger and food insecurity.[74]

Against that backdrop, sustainable agriculture isn't an option, it's a necessity. Right now, food production accounts for 70 percent of freshwater use. It's led to 80 percent of deforestation around the globe. And it's contributing an estimated 25 to 30 percent of global GHG emissions. It's the human endeavor most affected by higher temperatures, flooding, extreme weather events, and other impacts of climate change.

It's no exaggeration to say that today's food system is like the *Titanic*.

Immense.
Complex.
A marvel of engineering.
Thought to be invincible.
But racing to its destruction.

The difference, though, is that unlike the captain, the crew, and the passengers on the *Titanic*, we know that disaster awaits us if we don't change course—and do it fast. The amazing thing about growing food is that when it is done sustainably it can help mitigate climate change even as it strengthens food security, in developing and industrialized countries alike.

In Kenya, farmers who are working with the World Agroforestry Centre are growing nitrogen-fixing trees that provide a natural source of fertilizer for crops and, at the same time, eliminate the need for farmers to buy expensive fertilizer out of a bag.[75] That's a double win. Best of all, using these methods can increase yields by up to 300 percent.[76]

In Cambodia and across Asia, farmers are using the System of Rice Intensification (SRI) method that's helping family farmers conserve water resources, build soil health, and increase yields by up to 150 percent.[77]

Growing indigenous and traditional crops improves both incomes and nutrition while offering farmers a significant measure of insurance against crop failure and disease. These practices can often be replicated, adapted, scaled up, and applied on both small and large farms to improve water availability, increase crop diversity, improve soil quality, and mitigate climate change. How do we promote sustainable agriculture? We already know the ingredients.

We need government and NGOs to

- Invest in research and technical support
- Introduce nutritious diets
- Foster the use of environmentally friendly practices and appropriate technologies
- Provide training in management skills

But that's only part of the equation. We also need to stand with men and women who understand that sustainable agriculture can generate wealth, but to do so it needs consistent contributions from governments, from farmers and farmers' collectives and unions, from businesses, from the funding and donor communities, and from eaters themselves.

From Africa to Latin America to Asia, farmers are on the leading edge of the greatest agricultural transformation of our time. They're not scientists. Most never even finished school. They're separated not only by geography but by faiths and traditions that go back generations, long before any of us were born.

Despite their differences, they share a common vision for a sustainable and just food system. They can lead us in following the recipe for a better future through food.

VOICES FROM THE NEW FOOD MOVEMENT: Hilal Elver

Could you explain what the UN Special Rapporteur on the Right to Food does?

The right to food is one of the fundamental human rights legally protected by the UN International Covenant on Economic, Social and Cultural Rights, ratified by 164 countries. The Special Rapporteur monitors right-to-food violations, identifies general trends and policies related to the right to food, and undertakes country visits (at least two in a year), which provide the Special Rapporteur with a firsthand account of the situation concerning the rights to food in a specific country. Besides country reports, the SR presents two thematic reports per year to the UNHCR and UN General Assembly on selected priorities of the Special Rapporteur. The Special Rapporteur also communicates with states and other concerned parties with regard to alleged cases of violations of the right to food and other issues related to her mandate.

Prior to the 2015 climate talks in Paris, you said, "responding to the food demand through large-scale, production-oriented agricultural models is not the right solution. There is a need for a major shift from industrial agriculture to transformative systems, such as agroecology, that support the local food movement, protect small-holder farmers, and respect human rights, food democracy, and cultural traditions while maintaining environmental sustainability and facilitating a healthy diet." Now that the talks have ended, what progress do you feel we have made toward achieving these goals?

The Paris Agreement is the beginning of a new era for the next 15 years. At the same time, it is a continuation of the UNFCCC [UN Framework

Convention on Climate Change] climate change regime. It is the aftermath of the Kyoto Protocol, which was a binding period for the reduction of GHG emissions for countries that ratified it. So, we should consider the Paris Agreement a routine process of international climate diplomacy. Unfortunately, food systems have never been a popular subject in climate change negotiations, despite the very complex relationship between the two. What I defined in my report before the Paris meeting is still relevant: an ideal food system that is healthy for people, sustainable for the environment, and friendly toward climate change is the only viable system for the future. Now we have to wait and see whether states continue with business as usual or if they will shift their system towards the ideal one. Unfortunately, there is not much time for a trial-and-error period. If they commit to transforming their food and agricultural systems, among other policy changes such as to their energy systems, I hope they aim for what I defined in my report, which is commensurate with what many scientists, agronomists, food policymakers, and activists have been supporting for many years.

Moreover, the Paris Agreement gave only lip service to human rights, mentioning them only in the Preamble, rather than in operational articles. Needless to say, there is no "right-to-food" language anywhere in the Paris Agreement and the COP21 decision.

As a proponent of agroecology, you have said in the past that "agroecology incorporates local knowledge, technical knowledge, [and] traditional knowledge"—concepts that most people ideologically support—but that people doubt its effectiveness because of the persistent myth that increased production is the key to solving world hunger. How do you suggest we change the conversation from one of production to one of access and, in turn, promote agroecology?
We all know that hunger and malnutrition cannot be solved by pushing for more production-oriented policies. There is more than enough food in the world for all. The problem is accessibility and economic

inequality. Moreover, excessive production will bring us to a very dangerous dilemma regarding resource scarcity, loss of biological diversity, and, eventually, ecosystem failure. We still have challenges to believing in agroecology because we only ever hear one side of the argument. We need robust research and development funding to support agroecology on the one hand, and dissemination of this knowledge everywhere on the other. This is the only way to transform the agro-industry myth.

You have highlighted the necessity for women to hold land rights as a first step toward financially supporting small-scale farmers, who represent 70 to 80 percent of all farmers worldwide and who produce 70 percent of the world's food. What do you see as the remaining barriers to women's access to legally owning their land, and what steps are being taken to ensure women's access to land rights to achieve greater food security?
The role of women as food producers, providers, and consumers is enormous in food systems. Unfortunately, legal, economic, and cultural barriers block women from being effective players. Women's land rights, as well as access to other vital resources—such as to credit, to scientific and technical know-how, and, eventually, to markets—are very limited worldwide. Research shows that women's empowerment helps tremendously to eradicate hunger, poverty, and malnutrition, as well as economic development of families and countries as a whole. Therefore, countries should consider revising their legal systems as well as customary practices if they would like to deal with food security and economic development—especially in rural areas. If overall more than 60 percent of farmers are women, there is no question that reform should start with empowering women in agriculture and the food sector.

How would you respond to critics of programs that uphold the right to food but may unintentionally prolong conflicts like the ongoing one in Syria?

The Syrian war is a very complicated proxy war. A variety of countries, some in the region and some outside, have stakes in it. There are too many interests and too many sides. It is very hard to help civilians in such a tragedy, as parties in hot conflicts use food and water as a weapon of war, which is a grave violation of humanitarian law principles. Moreover, adding for-profit organizations to humanitarian aid can make such situations yet more complicated. The Syrian conflict started partly due to economic hardship in the aftermath of a long drought, and dysfunctional agricultural and subsidy policies. We see food crises are a triggering factor, but at the same time act as a weapon of war. At the end of the conflict, foreign aid might have a further negative impact on the local market, due to the dumping of cheap food from outside with which local producers cannot compete. So, food aid is very complicated, and it should be done only carefully, and for humanitarian purposes.

Hilal Elver is the third person to hold the position of UN Special Rapporteur on the Right to Food and a Global Distinguished Fellow at the Resnick Food Law and Policy Center of the University of California-Los Angeles School of Law. She is the author of *Peaceful Uses of International Rivers: The Euphrates and Tigris Rivers Dispute* and *The Headscarf Controversy: Secularism and Freedom of Religion*. Her most recent book is *Reimagining Climate Change*, co-edited with Paul Wapner and published in 2016.

Citation: "Interview with Hilal Elver: UN Special Rapporteur on the Right to Food," Food Tank (2016), https://foodtank.com/news/2016/06/interview-hilal-elver-special-rapporteur-right-to-food/. *Interview conducted by Kathryn Bryant in June 2016 and edited by Michael Peñuelas in August 2017.*

VOICES FROM THE NEW FOOD MOVEMENT: Hans R. Herren

What are the key challenges for agriculture sustainability now and in the future? What are the problems with the current situation?
The main challenges agriculture and the food system in general are facing are: How to eliminate the persistent nexus of hunger and poverty? How to deal with the nutrition and health issue? How to reduce inequities and cater for rural livelihoods?

The main problems agriculture is facing today are in the realm of adaptation to climate change; producing sufficient, diverse, and quality food, feed, and fiber at affordable prices while being remunerative for the producers and compatible with sustainable agricultural practices; the increasing competition from the biofuel sector; the increase of fossil energy prices and, in the medium and long term, also fossil energy scarcity.

Are there some agricultural production models that could help in achieving a higher level of sustainability?
Farmers and scientists have devised a number of agricultural practices over the years that are in line with the requirements of a sustainable and multifunctional agriculture, as requested in the IAASTD (International Assessment of Agricultural Knowledge, Science and Technology for Development) report. These go by different names, ranging from organic, biodynamic, agroecological, and low or zero tillage to conservation agriculture, with different levels of compliance to the sustainability and multifunctional goals.

The closest models to the set goals are agroecology and organic/ biodynamic agriculture, although even in these cases, more work is needed to meet social, environmental, and economic sustainability. In principle, there is a need to develop and build into these and new systems more resilience and regenerative potential, given that the present system still uses too much water and external, often nonrenewable inputs.

How is it possible to effectively manage the transition toward more sustainable production paradigms?

The transition from these unsatisfactory systems requires a new approach to research and extension that is participatory, localized, and includes the stakeholder beyond production, such as consumers/users, providers of inputs, and also the transformation and retail sectors. This is necessary, as production systems are shaped, in part at least, by these sectors that are beyond the farm gate and research lab sectors. There is also a need to recognize that agriculture and food are the responsibility of governments, and that these areas need major funding from the public sector, rather than being delegated to the private sector alone. The latter still has a large role to play—past the farm gate in particular—along the value chain from the farmer to the consumer. The transition will be further helped and supported by introducing true pricing of the products—i.e., include the production and transformation, as well as the indirect health costs externalities—into the retail price, removing all perverse subsidies and replacing them with payments for ecosystem services and rewards for sustainable practices.

Managing this transition will need political will and vision beyond what is presently experienced—at all levels of governance, from global to local—and new institutions to support and manage the paradigm change as well as a change in consumer and user behavior. It will also require a new systemic and holistic approach to analyzing the agriculture and food system, to identify the key leverage points and synergies to achieve the multifunctional agriculture goals while minimizing the negative feedbacks. New national agricultural policies will need to cater to the internal need of food, feed, and fiber production, as well as to the enabling conditions, which are just as important as rural infrastructure, access to markets, and both capital and insurances.

What kind of technology innovation and agricultural practices are required to meet the goals of sustainability in agriculture?

The main areas of knowledge, science, and technology needed to transition agriculture towards the sustainable systems required to address the above-mentioned challenges are rooted in the soil, so to speak! The world is facing many challenges, in particular the fact that in the developing countries, the soils have been largely mined of their nutrients, while in the developed countries, we have mostly overfertilized. The consequence of each practice are degraded, eroded, and low-fertility soils, devoid of the needed biota to assure sustainable fertility levels that allow quality and quantity production under the new stresses of climate change. Restoring soil fertility is therefore the number one concern, to which we need to add improved and more diverse cropping systems, with more different crops in the rotation; the inclusion of animals on farms; and new methods for pest and disease management that take advantage of the gifts of nature in the form of natural control mechanisms, either already built into plants through evolution or through system management practices that go from field to landscape scale.

What should be done to improve and promote agricultural best practices all over the world and further foster innovation?

It has been demonstrated in the 2011 UNEP [UN Environment Programme] Green Economy Report Agriculture that by implementing the basic tenets of sustainable agriculture as suggested in the IAASTD report, all key sustainability goals can be achieved, with investments that are below today's subsidy levels. The main factor being that agriculture needs to be green by design, rather than by making a few changes at the margins (green-washing), as suggested by most vested interest groups, including agribusiness. Investments need also to be made in enabling conditions, such as rural infrastructure, institutions, and along the value chain, to assure markets for agricultural products and to provide quality jobs in and around agriculture to keep the younger population in the rural areas

By making serious changes from agricultural sciences to political choices, agriculture and food systems can be made sustainable and able to deliver on the multifunctional goals—for the present and future—of meeting the food, feed, and fiber needs of a growing and more demanding population.

Hans Herren is a farmer, entomologist, the president and CEO of the Millennium Institute, and the president of the Biovision Foundation. He was also co-chair of the International Assessment of Agricultural Knowledge, Science and Technology for Development (IAASTD), which in 2008 published "Agriculture at a Crossroads," a report that assessed agricultural knowledge, science, and technology with respect to development and sustainability goals.

Citation: "Interview with Hans Herren," in *Eating Planet*, 2nd ed. (Milan: Edizioni Ambiente and the Barilla Center for Food & Nutrition, 2016), https://www.barillacfn.com/en/dissemination/eating_planet/. *Interview conducted by Tony Allan and edited by Michael Peñuelas in August 2017.*

VOICES FROM THE NEW FOOD MOVEMENT: Sieglinde Snapp

What initially inspired you to pursue a research career in applied soil ecology and to focus on southeast Africa?

I would say I was initially inspired by both passion and fate. I've always been interested in the African continent and wanting to help under-resourced farmers in Africa. My family also adopted African American kids, so we always sought to educate ourselves and the kids about the African side of their heritage. When I finished my Ph.D., I was hired by the Rockefeller Foundation for a position in Malawi, in Southern Africa. I have had the real privilege to continue working in that region ever since. I have students working in many parts of Africa, but I have maintained a particular focus on Malawi because of my personal connections to the country and the region, including my adopting kids from there. I consider myself very fortunate to have such a strong relationship with one place for so many years. Over two decades, now, I've really begun to understand everything from the politics to how the soils respond under smallholder agriculture.

What do you see as the most actionable opportunities for positively transforming food systems in southeast Africa? In other words, what gives you hope?

I'm inspired by the increasing rate at which people are developing the capacity to do things for themselves. In countries like Malawi, which was a dictatorship in 1993 around when I started working there, we now see a growth of civil society and a dramatic increase in education such that pretty much all children go to school. These long-term capacity changes are what really transform systems.

Within that larger context of education and civil society, we're helping to catalyze a significant and growing interest in crop diversity. My work is in trying to provide concrete alternatives to the monolithic kind of agricultural development that the Green Revolution adhered to. I like the concept that they use in India of a "Rainbow Revolution," where you include diverse crop options instead of a monolithic focus on grain. Agriculture should value diversity, including diverse ways to support soil building and a diversity of crops, that can help farmers cope with a rapidly changing world. There again is why education is so important: because farmers have to be able to harness diversity to adapt locally. Markets change constantly, climate variability is increasing, and the rate at which the world is changing is only accelerating.

Can you explain what it looks like to engage in the alternative research methods you champion, including participatory research and co-learning?

To me, process is as important as outcomes. I work to build capacity for farmers by providing options rather than one-size-fits-all silver bullets. I think that's the only real, lasting kind of agricultural transformation.

Whether we're talking conservation agriculture, organic agriculture, or whatever system, they all involve participatory practices and processes, whether or not they're labeled as such. Tilling equipment is a practical example. Organic farmers have to be really good at weed tillage, which means they constantly are experimenting with new tools and methods. The equipment that a community might have access to won't always be appropriate for all soil types, so they need to adapt and collaborate. Farmers have to literally weld their own solutions and interact with equipment dealers to source what they need. I don't think I know a small-scale farmer in America who isn't also a welder. Farmers live in participatory communities that prioritize co-learning, so adapting to do my research in a way that makes sense to the people I'm working with ends up producing better research.

In Malawi, I work on pigeonpea, a shrub that provides a lot of services in addition to food. It's basically "agroforestry-plus," because it does much more than produce a grain-like pulse. Pigeonpea is a tropical legume that is one of the main pulses grown in India and is increasingly grown throughout southern Africa. We think it's much more effective at improving soil organic matter than many other options because it grows into a shrub with a big root system and grows into the dry season, which means it captures more sunlight. Our focus on it came out of exploring, alongside farmers, what folks are trying in their fields. Farmers were testing different legumes in some of the regions we work in, and the farmers who were the most interested in soil fertility were using pigeonpea. They cut the bushes back at the end of the season and effectively turned it into a semi-perennial crop, which we agronomists didn't even know you could do! The agronomy recommendations for pigeonpea just talk about growing it as an annual, one-year crop in most regions, but these farmers had adapted it. They were getting a two-year crop before tilling it in. By co-learning, we were able to spread this innovation. Our research improved in quality because these farmers were doing experiments alongside us.

What do you see as the most significant challenges to your vision of a sustainable intensification-driven "greener revolution"?

I think one of our biggest challenges is to deal with our propensity to focus on silver bullet solutions. Wanting a one-size-fits-all answer precludes a focus on adaptation, on being willing to take the long-term view, on building local capacity, or on supporting farmers' own innovation.

Another challenge is that people come in assuming that fertilizers and both input and output markets are accessible everywhere. It's true that things like cereals do really well when there's fertilizer available, but they get you on a track where you have requirements you have to meet every year, like having a stable supply of fertilizer. Corn is especially nitrophilic, which means it's dependent on massive amounts of nitrogen being added, on top of its requirement for intensive weed management, which usually means herbicide. If you are only relying on one solution, and one that happens to be very fossil fuel dependent, then if you don't end up with access to those inputs at some point, you're trapped in a zero-sum game. If, on the other hand, you build diversity into the system instead, you automatically remove your reliance on these shaky systems and move onto a development trajectory that is more in keeping with Rainbow Revolution idea.

That said, lack of investment in infrastructure, including education, roads, and access to markets, is also a challenge. Since smallholder farmers are the main producers of food in countries like Malawi, a general lack of investment makes it very difficult for them to make a living, let alone to invest in protecting the natural resources like so many of them want to. Even when they're on the edge of survival or, as often happens in the U.S., are on the edge of losing their farms, our conventional system puts all the burden on the farmer.

A final challenge I'll mention is a rhetorical one: Our discourse is unbalanced. For example, people have this idea that organic farming is only for the elite. It's not just urban consumers who are poor, though. There is also a significant rural poor both right here in the developed world and across the rest of the world. One of the best ways to get poor people out of poverty is to give them enough money to support their agriculture. Many, many organic farmers have told me, "Look, with organic I finally have a system of farming that provides enough margin to pay my medical bills." Moreover, farmers are very much interested in a system that's less toxic. There's so much focus out there on genetically modified organisms (GMOs) while we ignore toxicity. We're growing more and more corn around the world and across southeastern Africa, for example, and that means more atrazine and other really toxic chemicals. Farmers want to protect their well water. We have a system where it's only farmers who get sick or who are about to lose their farms who feel pushed to take the risk to go organic. Here we have a way to address rural poverty, yet we say, "It's only for the elite." I think we need to reframe.

To support healthy agroecological systems in a developing country context, what do you think is the ideal balance between the roles of the major players, including governments, agroindustry, and traditional international development actors including multilateral funders like the World Bank, civil society organizations like One Acre Fund, and research groups like CGIAR?

I think we need a pretty equal balance because governments can get things wrong. I worry about countries going the route that Rwanda went. In Rwanda, the federal government decided to push for monoculture. The Ministry of Agriculture mandated that farmers could only grow a limited number of crops. The intention was to increase profits by having regions specialize in particular crops and becoming economies of scale. I thought peasant farmers would do what they wanted regardless, until the government started throwing people in jail. Farmers were forced to physically uproot the crops they were growing that didn't fit with the government plan. One of my students happened to be there, was caught in the middle of it, and wrote a paper on the situation. The country is now moving back on track, but the Rwandan story is a really important example of why you need balance amongst all of these partners.

Government needs to have a strong university system to provide a check and balance against bureaucrats and to provide education via extension services. Civil society actors need to be allowed to maintain a free press so that they can critique. Agroindustry is a very important part of this balance, as are the big research groups.

How can readers who are not directly employed by international research and development institutions productively engage with these issues?

I think the easy answer is to focus on advocacy not just for relief efforts such as direct feeding or food safety nets, but also for policy changes that can support governments becoming more democratic and open. We can support policy here in the U.S. that stops dumping food abroad and instead links it to investments in education and increased pressure for a stronger civil society and for strong constitutions. People need to be engaged with their legislatures and with NGOs that are on the broader side of supporting education rather than just filling immediate needs.

We also need to have more advocates for forms of agriculture that support the environment. We see a lot of talk about sustainable intensification, and I'm known for my work on it, but almost all of our plant breeding efforts and research efforts in the U.S. tend to still be on only the intensification side of the agenda. We're not thinking enough about, and haven't invested enough in, the environmental side. Researchers, universities, and CGIAR centers should be investing in the sustainability equation because no one else will do it. Unfortunately, in the end, we often get caught up in the short-term and the drive for modernization and for very simple systems because that is the model that looks easiest to replicate in the short run.

Dr. Sieglinde Snapp is a professor of soils and cropping system ecology at Michigan State University and associate director of the Center for Global Change and Earth Observations. Focused on understanding the principles of resilient cropping system design and biologically based soil management, she has pioneered the development of multipurpose crops including perennial wheat and perennial pigeonpea. Dr. Snapp is also well known for her extensive experience with participatory action research and her commitment to co-learning.

Citation: "Dr. Sieglinde Snapp Works Alongside Farmers towards a 'Greener Revolution,'" Food Tank (2017). *Interview conducted by Michael Peñuelas in July 2017.*

VOICES FROM THE NEW FOOD MOVEMENT: Vandana Shiva

What are the most important components of the food system that you would argue are broken and therefore deserve attention?
The fact that more than 815 million people are starving and more than 2 billion people are sick, and the fact that the planet is sick—with water disappearing, biodiversity disappearing, the climate damaged, and soils losing their fertility—are interconnected. They are connected by a model of farming that forgets the nutrition of the soil as it forgets the nutrition of people, a model that puts profits from extraction at its center.

This model means that small farmers can't feed themselves and join the ranks of the dispossessed. If they are able to keep farming, they're indebted, and they sell what they grow, keeping too little for themselves and their families. Of the 800 million people who are hungry, 500 million are farmers and food producers. A system that forgets that food is about nourishment ends up producing non-food. Non-food becomes junk food, and junk food creates all kinds of diseases.

This is the same model that is able to exploit water because it doesn't have to bear the cost of that exploitation. This model can push species to extinction. It can put 40 percent of the greenhouse gases into the atmosphere that are giving us climate change. Obsession with profits leads to destroying our food, destroying the Earth, destroying our farmers, and destroying our health.

Based on your experience, what approaches should developing countries take to prevent the problem getting worse?
I think that the most important point to highlight here is that so-called "developing countries" are called "developing" because we didn't

industrialize in the first industrial revolution. The large majority of people in our countries, even in China and India, are small farmers. In Africa and Latin America, that's definitely true. We need to treat our small farmers as our social capital because small farms produce more. If we start imitating the large-scale industrial corporate farming of the West, we will not only destroy our farmers, we will destroy our food security.

Secondly, because developing countries happen to lie in the part of the world that has higher biodiversity, we need to recognize that nature's capital, biodiversity, is real capital. We don't need capital in the form of financial loans from banks that are going to take away our land down the line. We don't need technologies that are already failing us, like genetic engineering. We need to have respect for the land, for our farmers, and for the older agricultural knowledge that has been time-tested. That is what the International Assessment of Agricultural Knowledge, Science and Technology for Development (IAASTD) report has pointed out. Neither the Green Revolution nor genetic engineering is a real solution to food security. Ecological farming, which is often linked to indigenous knowledge systems, is the way we can successfully increase production while conserving resources.

You argue that women have a specific role in this process. Can you explain what that role is and why you think it is important to both pay attention to and maintain?

Women have a specific role for two reasons. First, when we talk about the long history of the kind of agriculture that did not starve people, which did not create obesity, and which did not give us diabetes epidemics, it is a long history in which women had the knowledge and control. We need to turn back to women and ask, "How do we feed our people with nourishment?" That's why at Navdanya we run a grandmothers' university, so that we can learn once again how to treat food with respect.

Secondly, we have to recognize that this certain model of agriculture that is leaving 800 million people hungry and more than 2 billion obese is an agriculture with its roots in war. It is an agriculture that came out of war. Specifically, agri-chemicals came out of war. War has its roots in what I call the patriarchal mindset of man as dominator and man as a violent conqueror of the Earth and people. That model has become too heavy for the food system. We need the non-violence, the diversity, and the multifunctionality that women bring to agriculture.

You once said that "whoever controls our food system will control our democracy as well." Can you explain the logic behind that assertion?

At one level, it is the same thing the Secretary of State during the Nixon Administration, Henry Kissinger, said when he talked about food as a weapon. He said, in effect, "When you control weapons, you control governments and armies. When you control the food, you control people." In today's context, food is being controlled through the control of seeds. Monsanto has emerged as the single biggest player on the seed front. Sadly, the U.S. government, which has impoverished itself by outsourcing production, is now only collecting royalties from patented seed. This is taking away the democracy of the third world farmer to have their own seed and taking away the democracy of people worldwide to choose the food they grow and to know what's in that food. Food democracy in our times means having seed sovereignty and seed freedom. It means no patents on seeds. Having the ability to grow your own seed means you can grow your own food, and therefore defends the small farm. To ensure this, we also need to stop the perverse subsidies of US$400 billion that give industrial farming an unfair benefit. Third, we need to be much more aware of what we're eating and how it is grown. Democracy begins with food.

Vandana Shiva is an Indian scientist, author, and food activist. She started Navdanya, an organization and movement that protects the diversity and integrity of living resources, especially seed, and promotes organic farming and fair trade practices. She is one of the founders of the philosophy of ecofeminism, a founding board member of the Women Environment and Development Organization, and on the board of the International Forum on Globalization.

Citation: "Interview with Vandana Shiva," in *Eating Planet*, 2nd ed. (Milan: Edizioni Ambiente and the Barilla Center for Food & Nutrition, 2016), https://www.barillacfn.com/en/dissemination/eating_planet. *Interview edited by Michael Peñuelas in August 2017.*

CHAPTER 2

Food for Sustainable Growth

EATERS AND FARMERS ARE BEGINNING TO UNDERSTAND that the actions they take today can have long-lasting effects—both positive and negative—on a variety of social, economic, and environmental levels. A successful sustainable food system, one that not only meets the food needs of the population but also promotes public health and protects future generations, is paramount to ensuring a future for everyone, including young and old, rich and poor, farmers and eaters, policymakers and investors.

Food and agriculture are at the root of some of our most pressing environmental, health, economic, and social challenges. At the same time, they make up the sector that may hold the greatest opportunity for solving obesity, hunger, climate change, migration, and a whole range of other current and future problems. The sustainability of the agri-food chain depends not only on the commitment of farmers, businesses, and policymakers but also on the choices of individual eaters and families. Dietary decisions and personal actions can have a powerful effect on the entire socioeconomic landscape of food production and consumption.

In this chapter we explore the challenges the food system faces in the pursuit of sustainability and the actions being taken to reach that goal. We examine how our food choices simultaneously affect human health and the health of the environment. We explore the complexities of agriculture and identify how different farming practices can protect, and even improve, the environment. And we consider the pressures of feeding the world's ever-growing population as well as the negative ecological impacts made on air, water, and soil during the production and consumption of food.

For too many years, conversations about food have focused on quantity over quality, and the consequences have been disastrous: a world population that is simultaneously underfed and overfed and natural resources that, in many cases, have been severely depleted. Now, with the impacts of climate change becoming more evident, solving these problems is more urgent than ever before. This chapter will help us find the tools and best practices to encourage sustainable, long-lasting growth—and hope for a better food system.

The Food Pyramid Reimagined

The Double Pyramid model was developed in 2009 by researchers at the Barilla Center for Food & Nutrition (BCFN) Foundation to demonstrate the close relationship between what consumers grow and eat and the environmental impacts of producing these foods. Based largely on the Mediterranean diet, the Double Pyramid combines a traditional food pyramid with an upside-down environmental pyramid to provide consumers with more in-depth information about the foods they eat. In the environmental pyramid, foods are placed according to their ecological footprint—the amount of soil or water necessary to regenerate the resources used throughout the lifecycle of the products, from the field to the waste pile. The environmental pyramid is paired with a food pyramid representing the plant-based, nutrient-rich foods found in the Mediterranean diet.[1]

The Double Pyramid illustrates how the foods with the most nutrients are, more often than not, the foods that have the least environmental

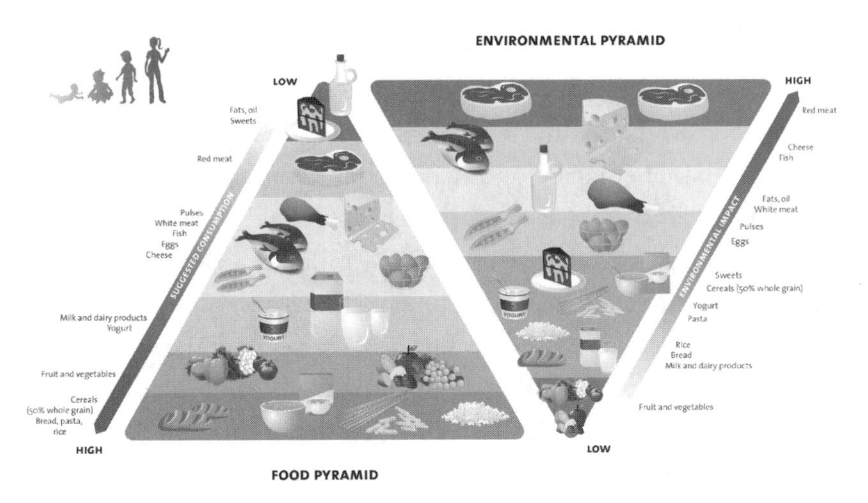

Figure 2.1. The Double Pyramid for growing children and adolescents. Source: BCFN, 2015, from p. 108 of *Eating Planet,* 2nd ed. (Milan: Edizioni Ambiente and the Barilla Center for Food & Nutrition, 2016).

impact. For example, fruits and vegetables, which are low in calories and high in nutrients, make up the base of the food pyramid. When we look at the environmental pyramid we see that the same foods tend to have the smallest ecological footprint. Vegetables occupy the bottom of the environmental pyramid, because the production of 2 pounds of them results in an ecological footprint of only 2 square meters. Two pounds of red meat, by comparison, has an ecological footprint of 54 square meters.[2]

Conversely, the foods with the greatest negative impact on human health—sugar, dairy, and meat produced in industrial animal operations, to name a few—are also the foods that create a wider range of environmental problems, from depleted soils to water pollution. Take industrial beef, for example. Not only is it high in cholesterol, the cattle raised for meat and milk are responsible for 65 percent of the livestock sector's greenhouse gas (GHG) emissions. Furthermore, livestock pasture uses

30 percent of Earth's entire land surface, and a third of this pasture is arable land that could be used to grow crops for humans.[3]

Luca Ruini, the health, safety, environment, and energy vice president at the Barilla company and an expert on environmental sustainability for the BCFN, explains the mission behind the Double Pyramid and the relationship between food and the environment. "Diet continues to be a key issue in the debate on the future of the planet, not least because of growing concern over the environmental impact from the production, distribution, and consumption of food," he says. "It is primarily in this direction that we are working, to conduct studies that correlate nutritional intake with environmental impact and also cost, which represents an increasingly important factor in consumer choices."[4]

Thanks to the graphic and intuitive simplicity of the Double Pyramid, it is also a powerful educational tool, capable of easily communicating the information acquired from studies on nutrition and the impact of food choices on the planet. It also has the ability to effectively educate eaters about the connection between diet and noncommunicable diseases, such as heart disease, type 2 diabetes, and some kinds of cancers.[5]

According to the World Health Organization, the number of people worldwide with diabetes rose from 108 million in 1980 to 422 million in 2014, and that total is predicted to reach 700 million by 2050. The global cost of this disease is approximately US$825 billion per year; the United States alone spends US$105 billion on the effort. If these trends continue, it is estimated that by 2050, 700 million adults around the world will have diabetes. Diabetes is often the result of a diet that is high in saturated fats and sugar and low in fruits, vegetables, and whole grains. Therefore, better education and better access to and affordability of more healthful foods can help eaters alter their diets and slow the trend in diabetes.[6]

In addition to creating consumer awareness, the Double Pyramid can be useful to farmers and businesses open to adopting more sustainable production practices. In 2016, the U.S. Department of Agriculture

(USDA) reported that farmers in the United States harvested 86.7 million acres of corn and 82.7 million acres of soybean. Despite these impressive totals, U.S.-grown corn and soybeans are rarely used to feed people. Instead, corn is used to make ethanol to fuel cars, and both corn and soybeans are fed to cattle, chickens, and pigs. Much of this is the result of U.S. crop subsidy policies, but if those policies were altered to reflect the need for more nutrient-dense, environmentally sustainable diets, it's possible those farmers could transform their practices and grow more diverse crops.[7]

Federal crop subsidies were introduced as temporary economic assistance for small farmers after the Great Depression, but today these government payments go largely to farmers who grow the crops considered most profitable. Proponents of these subsidies cite their contribution to keeping food prices stable; opponents counter that they undermine family farms by giving the most benefits to huge agricultural corporations, from industrial-scale farms to grain brokers and food processors.[8]

A study by the U.S. Centers for Disease Control and Prevention found that farm subsidies in the United States have contributed to obesity trends by making fattening and unhealthful foods both abundant and inexpensive. The study recommends changing farm subsidies to more strongly support farmers through trade regulations and higher consumer prices, policy changes that could discourage overconsumption and therefore reduce obesity and lead to healthier lifestyles.[9]

"[If] America is going to subsidize agriculture, the least it could do is subsidize healthy foods," says Richard Atkinson, a professor of medicine and nutritional sciences at the University of Wisconsin–Madison and president of the nonprofit American Obesity Association. "There are a lot of subsidies for the two things we should be limiting in our diet, which are sugar and fat, and there are not a lot of subsidies for broccoli and Brussels sprouts," he says. "What would happen if we took away the subsidies on the sugar and fat? Probably not much. They might go up a little bit, but the cost of the food is not the actual cost of the final products. But if we're trying to look for something political that might make a difference, try subsidizing fruit and vegetable growers so the cost is comparatively lower for better foods."[10]

The Double Pyramid can be a powerful tool on many levels. Environmentally and economically, it highlights the urgent need to both reduce the impact agriculture has on the global climate and fundamentally redesign the entire global agricultural system. Socially, the Double Pyramid can raise awareness among consumers and help them make more nutritionally sound, environmentally conscious food choices.[11]

The Double Pyramid is based on the Mediterranean diet (see Chapter 4) for two reasons. The first is the lower environmental impact made by the foods it recommends. The second is that the variety of foods it recommends—lots of fruit and vegetables, whole-grain cereals, olive oil, nuts, fish, and limited amounts of animal fats and meat, particularly red meat—are shown to promote health and longevity.

The Mediterranean diet is based on plant-based foods, rich in nutrients, including vitamins and minerals, and protective compounds, including fiber and antioxidants, that may promote health and reduce the effects of aging.

How Our Diets Affect the Environment

One of the biggest ways humans affect the environment is through food consumption and the production that supports that consumption. The combination of agriculture and related land use contributes at least 31 percent of total GHG emissions, higher than both energy (23.6 percent) and transport (18.5 percent). Livestock production, and meat in particular, is responsible for roughly 12 percent of overall carbon emissions, and dairy products contribute an additional 5 percent, according to the U.S. Environmental Protection Agency (EPA). In other words, it's clear that food choices play a key role in protecting our planet.[12]

The ecological footprints of the foods represented in the Double Pyramid are estimates of the amount of biologically productive land or ocean needed to provide the resources and absorb the emissions associated with a food production system; they're measured in square meters or global hectares. Further studies of the ecological footprints made specifically for water and air, as well as soil loss and degradation, can provide a more complete accounting of the environmental impacts on our food systems. It is important to note that these indicators are not the only impacts generated by the food chain (there are social and economic impacts as well), but they are certainly among the most significant in terms of real impact on the environment.[13]

Calculating ecological footprints can provide invaluable data—data that are meant not to frighten the world into inaction but to help farmers, consumers, and businesses develop more sustainable food systems that don't deplete natural resources or affect our health in negative ways.

Our Environmental *Food*prints

Carbon and water footprints are measures of the negative impacts made to our most important resources: air and water. The entire process of food production, from planting and growing to processing and packaging to transportation and preparation, affects the environment through both GHG emissions and the direct and indirect use of water.[14]

Dr. Andrew Lacis, an expert in climate and radiation studies at the Goddard Institute for Space Studies of the U.S. National Aeronautics and Space Administration (NASA), describes atmospheric carbon dioxide as the "thermostat regulating the temperature of Earth." A carbon footprint estimates how any particular production method affects the climate by measuring the method's GHG emissions. To measure a carbon footprint, emissions of all GHGs, including carbon dioxide, methane, and nitrous oxide, are converted into a common unit, called *carbon dioxide equivalent* (CO_2e), and added together. Because some GHGs

THE EVOLUTION OF

FOOD PYRAMID

FROM 1992 TO DATE

Mediterranean diet and other nutritional models around the world

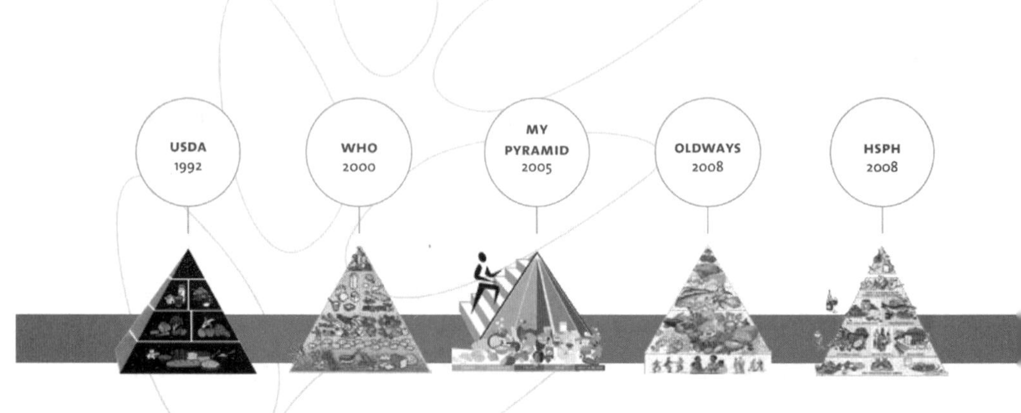

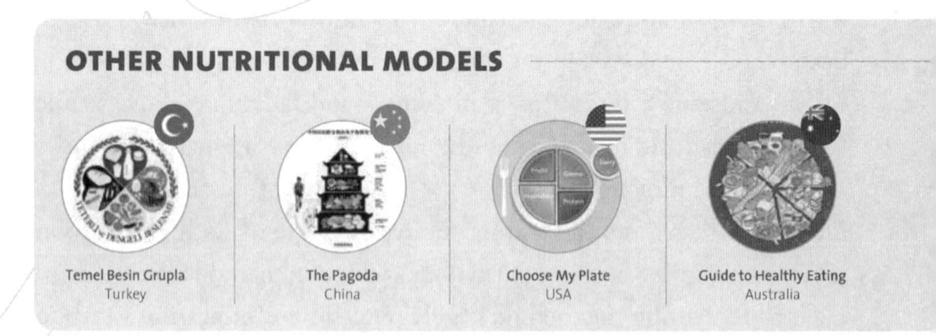

Figure 2.2. Evolution of the food pyramid, 1992–present, including the Mediterranean diet and other nutritional models around the world.

"The mediterranean diet represents a combination of skills, know-how, practices and traditions ranging from landscape to food, through cultivation, harvesting, fishing, preserving, processing, cooking and, particularly, consuming food."

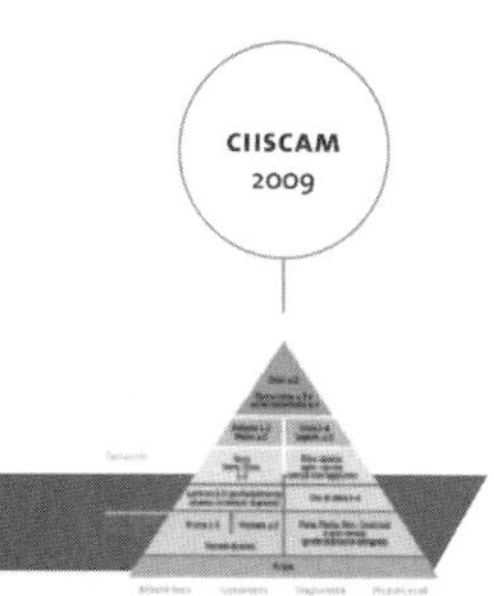

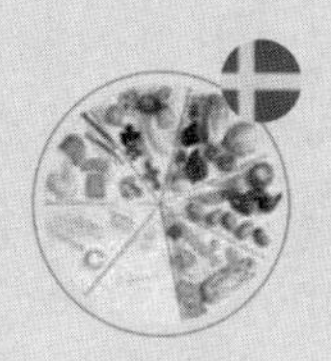

The Food Circle
Sweden

Food Bicycle
Korea

Food Spinning Top
Japan

The Food Rainbow
Canada

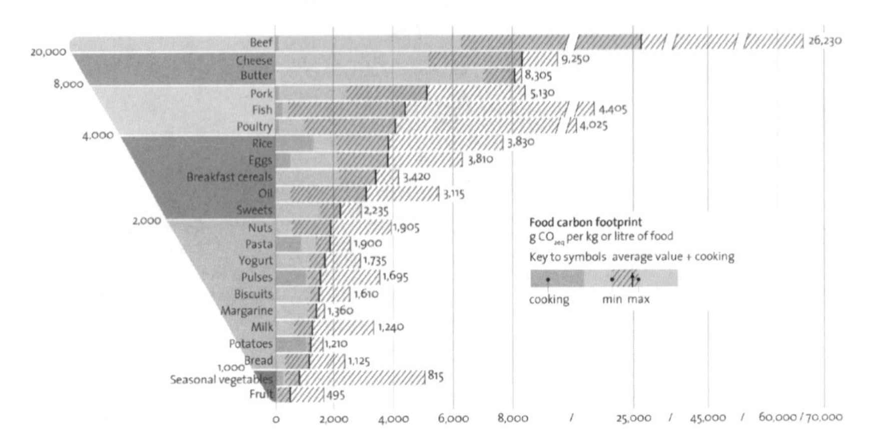

Figure 2.3. Carbon footprint of foods. The carbon footprint identifies GHG emissions responsible for climate change and is measured as a mass of CO_2eq. Source: BCFN Foundation, 2015, from p. 103 of *Eating Planet*, 2nd ed. (Milan: Edizioni Ambiente and the Barilla Center for Food & Nutrition, 2016).

are stronger than others, a measure of their individual strength, called global warming potential, is used for converting each to a CO_2e value.[15]

To put these numbers into context, beef production creates a carbon footprint of more than 20,000 grams of CO_2 per kilo. By contrast, a kilo of fish has a carbon footprint of slightly less than 4,500 grams, and poultry's is about 4,000 grams. The carbon footprint for beans and dried fruits is less than 2,000 grams of CO_2 per kilo produced, and vegetables and seasonal fruit create less than 1,000 grams per kilo produced.[16]

Recognizing the imminent peril posed by the buildup of GHGs in the atmosphere, 200 countries from around the world met in Paris in 2015 to sign a historic pact to fight climate change. In this pact, known as the Paris Agreement, participating countries agreed to limit CO_2 emissions enough to keep Earth's temperature rise to less than 2 degrees Celsius above preindustrial levels. Furthermore, they committed to reconvening every 5 years to set new emission reduction targets. A core part of the agreement was the creation of the UN Green Climate Fund, through which the world's wealthier, more industrialized countries would help poorer countries develop clean energy resources and adapt to the problems caused by climate change, including drought and rising sea levels. In a pledge made by the Obama administration, the United States would donate $3 billion to the fund by 2020.[17]

"The Agreement builds the case for both public and private actors to explore low-carbon and climate-resilient options. For developing countries, emerging economies, and their partners, the clear message is that growth without sustainability is off the table, whereas sustainable growth is a win for climate and development," said Barbara Buchner, head of the Climate Policy Initiative Europe. "For business and investors around the world, the Agreement means the direction of travel is clear, and with appropriate support, it is time to seize the opportunities on offer."[18]

However, the 2015 Emissions Gap Report, produced by the UN Environment Programme, illustrates that the combined current pledges from countries participating in the Paris Agreement actually put Earth on a course to warm by 3 degrees Celsius or more. If this temperature rise were to occur, it would further destabilize the ice sheets in Antarctica and Greenland, accelerate sea level rise, lead to more devastating droughts and heat waves, and cause a greater loss of coral reefs and other critical ecosystems.[19]

Riccardo Valentini, a co-recipient of the Nobel Peace Prize for his work with the Intergovernmental Panel on Climate Change, expressed concern about the terms of the Paris Agreement, not only for environmental reasons but for social and economic ones. "Agriculture is the third impacting sector for greenhouse gases after energy and transport, producing about 26 percent of the total human-induced GHG emissions in the atmosphere. Since 1990, greenhouse gas emissions from agriculture have soared by 20 percent and have doubled since 1960," he said. "In this scenario, maintaining global warming below 2 degrees Celsius would be highly improbable with such a high food demand and a traditional agribusiness as we know it."[20]

In May 2017, President Donald Trump announced the United States's withdrawal from the Paris Agreement, arguing that it posed a threat to the American economy. This withdrawal (which will take up to 4 years to complete) will probably short-circuit the promised United States contribution to the UN Green Climate Fund to help developing nations curb their carbon emissions, which some think may discourage these nations from continuing their fight against climate change. (Although the full US$3 billion promised by the United States appears to be in real jeopardy, before President Obama left office the United States had already contributed US$1 billion to the fund.)[21]

Critics were appalled by the withdrawal and argued that the pact was no threat to the U.S. economy. Senate minority leader Charles E. Schumer (D–NY) called Trump's decision "a devastating failure of historic proportions. Future generations will look back on President Trump's decision as one of the worst policy moves made in the 21st century because of the huge damage to our economy, our environment, and our geopolitical standing."[22]

Former president Barack Obama echoed Schumer's sentiments, but he maintained a more hopeful tone. "The nations that remain in the Paris Agreement will be the nations that reap the benefits in jobs and industries created. I believe the United States should be at the front of the pack," he said. "But even in the absence of American leadership, even as this administration joins a small handful of nations that reject the future, I'm confident that our states, cities, and businesses will step up and do even more to lead the way, and help protect for future generations the one planet we've got."[23]

In May 2017, Obama attended the Seeds and Chips Food Conference in Milan, Italy, which focused on the ways technology can aid food production. The former president explained that the rise in meat consumption around the world has caused an increase in carbon emissions and that our current food system is not sustainable in the face of climate change.

In the conference's keynote speech, Obama noted, "Even if every country somehow puts the brakes on the emissions that exist today, climate change would still have impact on our world for years to come. . . . We've already seen shrinking yields and spiking food prices that in some cases are leading to political instability." When asked why changing people's minds about sustainable food choices is such a challenge, he replied, "Because food is so close to us and is part of our family and is part of what we do every single day, people, I think, are more resistant to the idea of government or bureaucrats telling them what to eat, how to eat, and how to grow."[24]

Although food and farming are major instigators of climate change, they can also be part of the solution. Sustainable food production and consumption practices can help mitigate and even reverse some of the impacts of climate change. This idea is supported by Anna Lappé, author and co-founder of the Small Planet Institute and director of the Real Food Media Project, who states that "food is a culprit in climate change and also part of the solution. Casualty, culprit, cure—the three C's. We need to help people see that farmers are on the front lines of the crisis; they are among those most impacted. We need to develop policy and action to help farmers who are suffering and to help them develop new strategies that reduce their impact and improve their resilience. We also want people to see that farmers aren't the adversaries, but really are the source of potential positive change."[25]

Farmers are creating that positive change in a multitude of ways. One way is through carbon sequestration, including techniques such as cover cropping and agroforestry that literally sequester carbon in the soil, which can increase the sustainability of food systems while decreasing the negative effects of agriculture on climate.[26]

A report by the Rodale Institute suggests that "recent data from farming systems and pasture trials around the globe show that we could sequester more than 100 percent of current annual CO_2 emissions with a switch to widely available and inexpensive organic management practices, which we term 'regenerative organic agriculture.' These practices work to maximize carbon fixation while minimizing the loss of that carbon once returned to the soil, reversing the greenhouse effect." Mainstream adaptations of geoengineering techniques similar to carbon sequestration have the ability to mitigate climate change and combat its most harmful effects before it is too late.[27]

The carbon footprint is an informational tool consumers and policymakers can use to monitor changes to the environment caused by GHG emissions. Similarly, the water footprint is a tool that measures how much water is consumed, both directly and indirectly, and polluted during the entire cycle of a food's production. The data used to calculate a food's water footprint offers a more in-depth understanding of how limited freshwater resources are being used, allowing consumers to alter their consumption habits to be more responsible.[28]

The water footprint calculates the volume of freshwater used to produce food. It considers not only the amount and type of water source

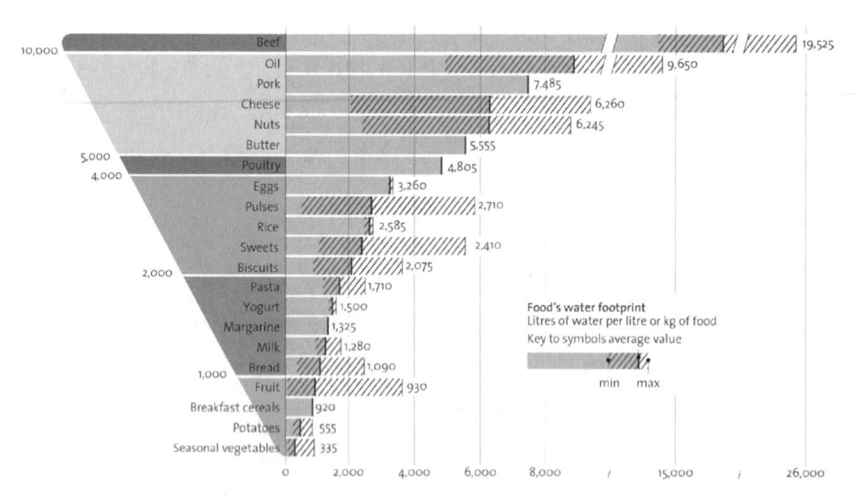

Figure 2.4. Water footprint of foods (liters of water per liter or kilogram of food). Source: BCFN Foundation, 2015, from p. 103 of *Eating Planet*, 2nd ed. (Milan: Edizioni Ambiente and the Barilla Center for Food & Nutrition, 2016).

used or polluted but also the use of "virtual water," which indicates that the water was removed from one place through agricultural production and moved elsewhere.

Until recently, water scarcity was typically thought of as a problem for underdeveloped countries, but water shortages are now being felt worldwide. The combination of population growth, global warming, and changing food preferences will exert growing pressure on water sources for agricultural use in the coming decades, making this an issue to be addressed immediately.

Overall, about 70 percent of the planet's limited freshwater supply is used for agriculture. Low- and medium-income countries use up to 95 percent of their water for farming; in more developed industrialized countries, large amounts, nearly 60 percent, are used in industry. For example, Malaysia creates a water footprint of 243 cubic meters per ton when growing oil palm, which the country produces at a rate of 90 million tons per year. By comparison, the United States creates a water footprint of 3.25 cubic meters per typical 2-ton car, which it builds at the rate of 60 million per year.[29]

According to the World Health Organization, water scarcity will affect half of the world's population by 2025. Areas using the larger share (greater than 20 percent) of available resources will feel the effects of water scarcity more acutely. The shortages are expected to expand to the United States, Continental Europe, and southern Asia and to worsen significantly in areas of Africa and the Indian peninsula.[30]

"We are facing a very complex situation globally where we have countries with plentiful water supplies who use it badly and others who are forced to resort to unsafe water," explains Marta Antonelli, research program manager at the BCFN. "We know that much of our water consumption goes toward food production, and as the world's population is set to grow to 8.5 billion by 2030 and almost 10 billion by 2050, it is clear that feeding this number of people will necessitate a further increase in water consumption of at least 20 percent (in the best-case scenario). Depending on consumption, it could be over 50 percent by 2030 and above 70 percent by 2050."[31]

Every day, the average person consumes at least 2 liters of water. Without realizing it, he or she also uses up to 5,000 liters of "invisible" water, hidden in food and food production. If eaters adopted a vegetarian diet, however, the per-person daily consumption of virtual water could fall to 1,500 to 2,600 liters as opposed to the 4,000 to 5,400 liters consumed in a meat-heavy diet. In practical terms, this means that, for example, eating a bowl of chickpea soup with a plate of green beans and steamed potatoes with grated parmesan and a portion of fruit also means eating, without realizing it, 1,446 liters of water. Meanwhile, replacing that meal with one of a steak, a mixed salad with olive oil, a slice of bread, and a portion of fruit increases water consumption to 3,244 liters. The volume of hidden water in a meal is significantly lower in a vegan diet; for example, one serving of vegetable minestrone with pasta, a portion of hummus, and a slice of bread contains "only" 940 liters of water.

Consumers have more power than they realize to make a difference through both small and big changes in their diets. Following the Double Pyramid is one way to achieve better health for both eaters and the planet, as is supporting more sustainable agricultural practices that help conserve water.

The Water Economy: How Much Do We Have?

Water is so precious because it is limited by nature. The science of water economy studies the way in which water resources are limited and how they must be managed to satisfy farming needs without creating social inequalities and unsustainable environmental impacts.

Overall, the planet possesses some 1.4 billion cubic kilometers of water. However, it is estimated that less than 45,000 cubic kilometers (0.003 percent of the total) is theoretically usable and that only 9,000

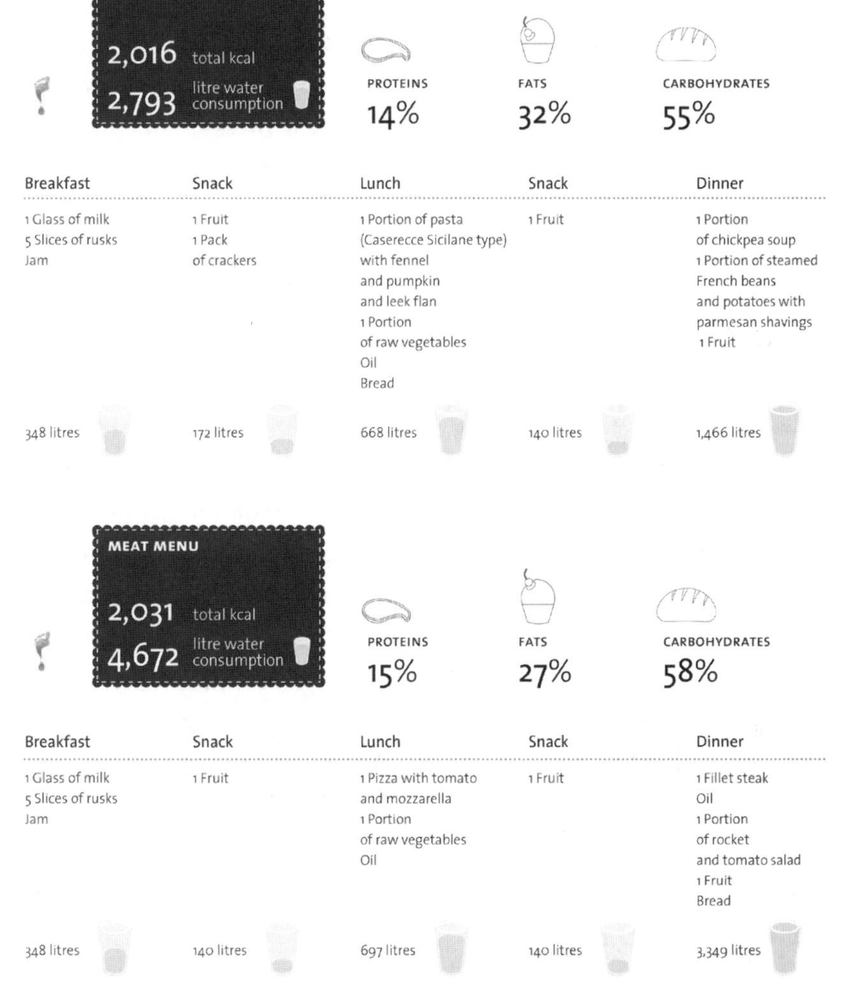

Figure 2.5. Virtual water consumption and eating habits: two menus compared. Source: BCFN Foundation, 2015, from p. 137 of *Eating Planet*, 2nd ed. (Milan: Edizioni Ambiente and the Barilla Center for Food & Nutrition, 2016).

to 14,000 cubic kilometers (approximately 0.001 percent of the total) is suitable for human use, which means it is of adequate quality and is accessible at an acceptable cost.[32]

Freshwater resources are distributed unequally across the globe. According to "The World's Water," a report updated every 2 years by the Pacific Institute, nearly 65 percent of the world's drinking water is located in just 13 countries: Brazil (14.9 percent), Russia (8.2 percent), Canada (6 percent), the United States (5.6 percent), Indonesia (5.2 percent), China (5.1 percent), Colombia (3.9 percent), India (3.5 percent), Peru (3.5 percent), Congo (2.3 percent), Venezuela (2.2 percent), Bangladesh (2.2 percent), and Burma (1.9 percent). On the other hand, a growing number of countries are facing grave water shortages, and some are even looking at annual per capita availability of less than 1,000 cubic meters.[33]

On a global average, the World Health Organization estimates, 842,000 diarrheal deaths occur each year; 361,000 of those deaths are of children less than 5 years old who died because of unsafe drinking water. According to UNICEF, 768 million people worldwide lacked access to safe drinking water in 2015; one in six people do not reach the minimum standard set out by the United Nations of 20 to 50 liters of freshwater per person per day.[34]

With statistics like these in mind, in 2010 the United Nations recognized the "right to water" as a fundamental and essential human right. This right establishes that everyone, without discrimination, has the right of access—physically and economically—to a sufficient amount of water that is safe to drink.[35]

This recognition of water as a basic human right was proclaimed in the UN General Assembly's Resolution 64/292. In response to the resolution, the UN Human Rights Council directed member states to "develop appropriate tools and mechanisms, which may encompass legislation, comprehensive plans, and strategies for the sector, including financial ones, to achieve progressively the full realization of human rights obligations related to access to safe drinking water and sanitation, including in currently unserved and underserved areas."[36]

Managing Our Supply: The "Virtual Water" Trade and Water Privatization

Water scarcity can be a source of conflict between those with a sparse supply and those with plenty, so the fair and careful monitoring of supply management and distribution is of global importance. In addition to the water used for drinking and agriculture, virtual water—water used during the process of worldwide trade—is an important resource that must be quantified and analyzed.

The concept of virtual water was introduced by Tony Allan, one of the world's leading experts on water. Allan defines virtual water as a means to "reveal the hidden factors of our real global water consumption." He also describes the urgent need to promote this concept. "Already, our over-consumption and mismanagement of water has had a very serious impact on our water environments and the essential services they provide. . . . Most of us don't have the slightest idea about the sheer volumes of water involved in our daily lives. To make a cup of coffee, it takes 140 liters. That's the true amount of water used in growing, producing, packaging, and shipping the beans you use to make your morning coffee," Allan says. He thinks the use of virtual water was less of a concern in the past because "the ratio of water to people was so massive that it was as if our water supply was infinite." Now, he says, "it is not. And now, with a global population pushing seven billion, water scarcity is not just a possibility. It is already a reality for many."[37]

Virtual water is traded in huge volumes as crops that need large amounts of water to cultivate are shipped far and wide, not always with sensible results. For example, three of the world's top 10 wheat-exporting countries are seriously short of water, and three of the top 10 wheat importers are blessed with an abundance of it. The level of interdependence between countries in the virtual exchange of water resources is critical, however, and it is destined to grow in the future, given the ongoing, often controversial deregulation of international trade.[38]

Water trade expert Dennis Wichelns, a professor of economics and executive director of the Rivers Institute at Hanover College, analyzed the trade patterns between Jordan and other countries, including the United States. Because Jordan has little water, it trades with other countries for commodities that use a lot of water to produce. Wichelns explains that "[Jordan] imports five to seven billion cubic meters of water in virtual form per year, which is in sharp contrast with the 1 billion [cubic meters] of water withdrawn annually from domestic water sources." Therefore, he states, "People in Jordan survive owing to the fact that their 'water footprint' has largely been externalized to other parts of the world, for example the U.S."[39]

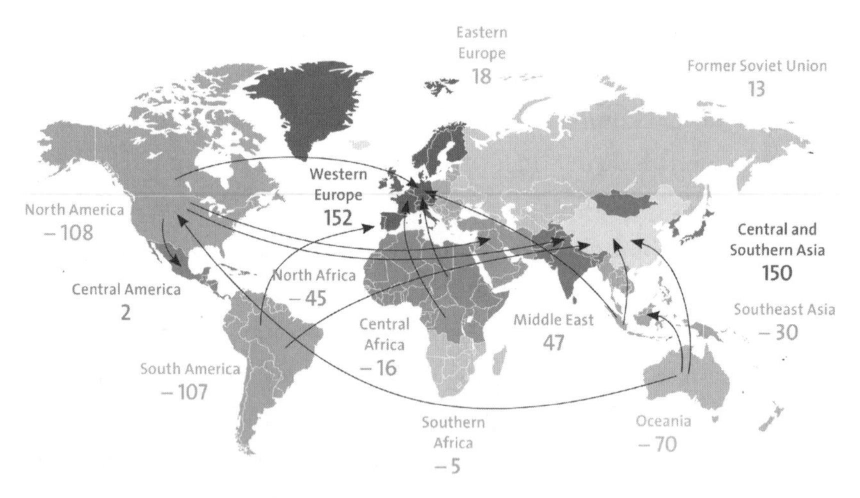

Figure 2.6. Virtual water trade. Patterns and volumes of the global trade in virtual water (in Gm3/year) embodied in agricultural products (negative numbers indicate net virtual water importers). Source: A.Y. Hoekstra, "Water Neutral: Reducing and Offsetting the Impacts of Water Footprints," UNESCO-IHE Institute for Water Education (2008), http://waterfootprint.org/media/downloads/Report28-WaterNeutral.pdf

If demand grows and resources dwindle—in part because of pollution and climate change—then clearly the economic value of water will grow, and the gap between those who have plenty of water and those who do not will provoke new conflicts. Water privatization—when private sectors purchase the right to participate in the sanitation and distribution of water resources—has been cited as a possible, albeit divisive solution to this problem.[40]

Opponents of water privatization point out how risky it is to entrust the management of water resources to private entities. The greatest of those risks, they say, is being sure that water managers respect their obligation to develop the water supply in poorer areas, where consumption is lower.[41]

This risk is playing out in South Asia, in the tensions between India and Bangladesh over use of the water in the Ganges–Brahmaputra Delta. Bangladesh, located farther downstream from the delta and the more economically and politically disadvantaged country, contends that water allocation and privatization favor India, despite the water-sharing treaties the countries have signed. This conflict raises ethical concerns, because any time one country is "cheated" out of its full water share, the less water it has to distribute to its citizens, which obviously worsens water insecurities.[42]

Supporters of water privatization point to how much more efficient the private sector is at managing water than the public sector. Outsourcing water management to private entities, they say, could improve distribution and make it possible to divide maintenance costs between different companies, resulting in lower costs to consumers.[43]

Private companies, such as the American States Water Company and its utility subsidiary, Golden State Water, work together to divide costs for roughly 260,000 consumers in California. Dividing the costs of utility and electric services may help save consumers from having to pay those costs themselves, were the company publicly owned.[44]

Private companies may also focus on developing effective distribution monitoring systems. One company, Environmental Health and Safety

Support, does so through its supervision of groundwater well designs and its evaluations of aquifer recharge.[45]

Above all, the most effective monitoring systems are those that manage water with citizens' interests in mind. Knowledgeable, experienced farmers, whom Tony Allan refers to as the de facto managers of the world's water, are a vital component of realizing a fair system. The importance of farmers' roles further underscores the symbiotic relationship between agriculture, food, and the environment.[46]

Soil Loss and Degradation

The urgent need to responsibly manage the care and quality of Earth's soil is every bit as important as managing its limited water resources. In fact, soil is one of agriculture's most important inputs.

Nearly 40 percent of all land on Earth is used for activities relating to agriculture and livestock, with 4.4 billion hectares of terrain suitable for farming. Alarmingly, in the past 40 years, 30 percent of once-arable land has become unproductive. According to the World Wildlife Fund, about 30 percent of total global land area is degraded, and topsoil is being lost at a rate of about 10 million hectares per year. Brazil, one of the largest agricultural producers in the world, loses 55 million tons of topsoil every year to erosion from soy production alone.[47]

A corresponding study by the International Food Policy Research Institute confirms these statistics, reporting that roughly 3 billion people around the world reside in degraded lands. In the developing regions of sub-Saharan Africa, South America, and Southeast Asia, problems related to soil quality affect more than half of the acreage being cultivated.[48]

Economically disadvantaged countries vulnerable to soil degradation face devastating impacts as the effects of climate change worsen. For small island states that rely heavily on agriculture, such as the Maldives and Samoa, rising sea levels induced by climate change will result in a loss of land, inhabitant displacement, increased soil erosion, and soil

Box 2.1. Confronting Climate Change on Behalf of the Future

The planet's climate is changing at an unprecedented rate. In recent years, with the aim of reducing the emissions of GHGs and of mitigating the effects of climate change, the world's nations have signed a number of international treaties, from the 1992 UN Convention on Climate Change to the 1997 Kyoto Protocol to the 2015 Paris Agreement.

We are now at a particularly delicate moment for the health of our planet. If the growth in food production does not include a substantial commitment to the use of responsible methods, it could result in an 80 percent increase in emissions produced by agriculture and in added pressure on water, land, and other natural resources.

Many countries are developing strategies to reduce their GHG emissions, including those from food production. Agriculture and related land use changes represent about a third of total emissions. According to the Food and Agriculture Organization of the United Nations, agriculture is the main cause of deforestation, loss of biodiversity, and soil degradation. Food loss and waste are also large contributors to GHG emissions.

The food industry's role in the fight against climate change is crucial, and improving agro-food systems is essential to the effort. It is equally important that consumers change their eating habits. It has been reported that if the world population adopted a diet of about 2,100 kilocalories per day, and if only 160 of those calories came from meat, it would result in a reduction of approximately 15 gigatons of carbon dioxide emissions annually, or about one-third of the GHGs emitted globally in 2011.

Through their production processes, the food and beverage sector accounts for roughly 30 percent of the total impact made to the environment, compared with 15 percent for the transport sector. If only the emissions of GHGs were considered, food would make the greatest contribution to climate change. Meat consumption is particularly relevant, because its production accounts for 12 percent of total emissions (dairy products account for an additional 5 percent or so).

Tara Garnett, head of the Food Climate Research Network, recognizes the impacts made by food choices. "Broadly speaking, eating fewer meat and dairy products and consuming more plant foods in their place is probably the single most helpful behavioral shift one can make to reduce food-related greenhouse gas emissions," she says. She underscores how crucial food choices are to the preservation of our planet. Eaters, she says, have the power to make the important food choices, and an increase in their awareness is paramount.[a]

a. A. Tukker and B. Jansen, "Environmental Impacts of Products: A Detailed Review of Studies," *Journal of Industrial Ecology* 10, no. 3 (2006), http://onlinelibrary.wiley.com/doi/10.1162/jiec.2006.10.3.159/full; "Prosperous Living for the World in 2050: Insights from the Global Calculator," Climate-KIC and International Energy Agency (2015), www.globalcalculator.org/sites/default/files/Prosperous%20living%20for%20the%20world%20in%202050%20-%20insights%20from%20the%20Global%20Calculator_0.pdf; T. Garnett, "The Food Sustainability Challenge," Food Climate Research Network (2014), https://www.sbs.ox.ac.uk/sites/default/files/Skoll_Centre/Docs/essay-garnett.pdf; B. Bajzelj, K. Richards, J. Allwood, P. Smith, J. Dennis, E. Curmi, and C. Gilligan, "Importance of Food-Demand Management for Climate Mitigation," *Nature Climate Change* 4 (2014): 924–9, https://www.nature.com/articles/nclimate2353; R. Oppenlander, *Food Choice and Sustainability: Why Buying Local, Eating Less Meat, and Taking Baby Steps Won't Work*, Hillcrest Publishing Group (2013).

salinization. The amount of water available for agriculture will also decrease as saltwater intrusion increases. The decrease in cultivated land availability and the increase in soil salinization will prove devastating for exports and will increase the fragility of these island ecosystems.[49]

Currently, 25 percent of Earth's soil is severely damaged, and only 10 percent shows any sign of improvement. Furthermore, in the past 150 years half of the planet's top layer of soil has been lost. There are many contributing factors. Deforestation, for example, has caused rapid erosion on the Loess Plateau in China. In the western United States, overgrazing has reduced topsoil depth and caused desertification. In India, intensive farming and the repeated cutting of trees have caused the fertility of the soil to plummet, threatening, among other things, the growth of wild medicinal plants.[50]

Soil loss and degradation is potentially one of the major threats to food systems across the world. Sustainable agriculture is more than an option; it is a necessity if we are to combat hunger, poverty, ecolo-gical degradation, food waste, and other problems. Respect for diversity, both biological and cultural, and the rediscovery of traditional practices can help us find ways to grow crops that are resistant to pests, diseases, drought, floods, and other disasters that will probably increase with climate change.[51]

In general, the planet has a finite amount of arable agricultural land, and what it has is quickly running out. A large part of it is already farmed or used in other ways (currently 1.6 billion hectares are cultivated, of which only 20 percent—300 million hectares—is on land marginally adapted for agriculture) or is threatened by erosion or overfarming.[52]

As the effects of climate change and overfarming increase, soil degradation is having a large impact on agriculture worldwide. In the United States, the average rate of soil erosion on tilled land is 7 tons per acre, per year, and this degradation decreases productivity by an estimated US$37.6 billion a year. In the American Midwest specifically, which contributes the largest share of globally traded corn and soybeans, yields for those crops are expected to drop by almost a quarter.[53]

Costa Rica loses 860 million tons of tillable soil each year to erosion and degradation, and Madagascar loses 400 tons per hectare of its distinctive red soil, which turns the island's rivers and the inshore waters of the Indian Ocean blood red. Côte d'Ivoire offers dramatic evidence of the dangers of deforestation. In the West African country, bare slopes lose 138 tons of soil per hectare each year, and cultivated slopes lose 90 tons of soil per hectare each year. By contrast, forested slopes lose just 0.03 tons of soil per hectare each year. In the Amazon rainforest, a football field–size piece of forested area is clear-cut each second, and habitats are destroyed by the resulting soil erosion and desiccation. Logging, particularly clear-cutting, is the leading cause of soil erosion worldwide, and almost half of the world's topsoil has been lost due to erosion.[54]

Although the statistics may seem overwhelming, farmers and eaters alike have enormous power to reverse damage to land, air, and water by making more responsible production and consumption choices. Fortunately, there are hundreds of organizations working to fight soil degradation worldwide. The nonprofit Savory Institute is working internationally to restore depleted grasslands. The institute promotes what it calls holistic management, a method of managing livestock that has been proven to help restore grasslands to their natural state. The organization is currently working in Africa, North and South America, and Turkey.[55]

Prolinnova is a nongovernment organization and an international multi-stakeholder network that promotes local ecologically innovative processes to manage agriculture and natural resources. The organization puts on farmers' fairs at local, national, and even international levels. For example, a national farmer innovation fair was held in Nepal in 2009, a regional fair for East Africa was held in Kenya in 2013, and a West African Farmer Innovation fair was held in Burkina Faso in 2015.[56]

An organization called A Growing Culture is building the Library for Food Sovereignty to connect farmers from around the world—from the Himalayas to the Great Rift Valley to the Mekong Delta—and show them how they can do more sustainable food production. The Library for Food Sovereignty will enable farmers to compare ideas and approaches and learn from one another's successes and failures. It will greatly expand farmer access to new ideas and innovative practices from around the world.[57]

The One Acre Fund is promoting climate-smart farming techniques to help farmers adapt to and eventually begin to mitigate the effects of climate change. Farmers who enroll with the One Acre Fund can choose from a range of crops and seed varieties, and the group's continuous research aims to optimize its recommendations on what plants perform best at the local level, including in drought-prone areas. The One Acre Fund also offers training in techniques that help keep soil resilient and rich in nutrients, including crop rotation, crop diversity, and composting. In addition, the organization is supplying farmers with tree seeds or seedlings, and as a result, farmers have planted millions of trees in the past 5 years.[58]

Food Loss and Food Waste

Despite the efforts made by organizations to combat the negative impact of soil degradation, food loss and food waste continue to hurt not only the environment but the lives of the hungry as well. Food waste tends to be insidious: A little bit is lost in the field, a bit more is lost in storage, a bit then gets lost in transit, and, finally, a small percentage is lost at restaurants and homes and retailers. Put it all together, though, and the waste is significant.

At least 1.3 billion tons of food is lost or wasted every year, in industrialized and developing countries alike. Some 222 million tons of food is wasted annually in industrialized countries, including the United States, where roughly one-third of food is thrown away by consumers and retailers for reasons like the misinterpretation of expiration and sell-by dates. In sub-Saharan Africa and other parts of the developing world, an equal amount of food is lost because of poor infrastructure and pests and disease. As a result, much of the hard work that farmers do to grow crops goes to waste, driving them deeper into poverty.[59]

According to the FAO, wasted food costs some US$680 billion in industrialized countries and US$310 billion in developing countries. Meanwhile, in 2016, 108 million people were reported to be facing crisis-level food insecurity or worse.[60]

Although food waste presents obvious moral and economic dilemmas, it also creates environmental problems. As food decomposes in landfills, it releases methane, a GHG that is 27 times more potent than carbon dioxide. According to the EPA, landfills are the third-largest source of human-related methane emissions in the United States. Therefore, combating the issue of food waste can not only help strengthen food security and provide economic benefits, it can also improve the efficiency of natural resource use and reduce environmental impacts.[61]

The good news is that preventing food waste can be both simple and inexpensive, and some of the most interesting and effective innovations are taking place in the developing world.

Organizations across the globe are also coming together to combat this problem. The One Acre Fund is helping farmers in East Africa learn not only how to store their crops better but also how to keep track of what they're growing and how much is lost. They are using simple tracking sheets to see how much they grow and how much is saved from one season to the next, and these farmers are also using better storage bags, which prevent pests from destroying crops.

In the United States, Second Harvest has become a pioneer in collecting food that otherwise would have been wasted and distributing it to homeless shelters and low-income families. The Food Recovery Network, another trailblazer, has mobilized hundreds of college students to collect perishable food from their campuses and communities and distribute it to those in need.

Of course, on the consumer side, the solutions to limiting food waste are simple: Don't buy more than you can eat. Store your produce properly. Take leftovers home. Don't throw food away that hasn't gone bad. And trust your senses, not expiration dates, to tell you whether food is safe to eat.

"Use by," "sell by," "best by," and "expires on" labels are not federally regulated, and both consumers and retailers are throwing away food unnecessarily. In 2016, this needless waste prompted U.S. senator Richard Blumenthal and congresswoman Chellie Pingree to introduce the Food Date Labeling Act, which would standardize food date labeling. The bill would require the nationwide use of uniform food date label terminology (i.e., "sell by") that is both accurate and clear to consumers. The aim of the bill is not only to reduce consumer confusion but also to simplify regulatory compliance for companies and cut supply chain and consumer waste of food and money. Additionally, the bill would ensure that food could be legally sold or donated after its quality date and would educate consumers about the meaning of the new labels so that they could make better economic and safety decisions.[62]

Still, action on the consumer side doesn't absolve food manufacturers and food retailers of their responsibilities. In 2015, federal agencies, including the USDA and the EPA, issued a report that, among other things, set a goal of reducing food waste by half by 2030. In line with this government goal, large private companies, including Kellogg and PepsiCo, pledged to meet the same target. The report points out that the 2018 farm bill will present Congress with an opportunity to prioritize the issue of food waste and identifies four areas of focus: food waste reduction, food recovery and redistribution, food waste recycling, and food waste reduction administration. According to the report, the farm bill "provides an appropriate vehicle for the federal government to take concerted action against food waste" because it affects most areas of our food system.[63]

Recently, for the first time ever, a global standard for what constitutes food waste and guidelines for measuring, reporting on, and managing that waste were released by the Food Loss and Waste Protocol Partnership, led by the World Resources Institute, in hopes that the guidelines would help minimize food loss and waste. The data surrounding food loss and waste vary widely from country to country, which can make it difficult to compare situations and develop solutions. The Food Loss and Waste standard provides consistent accounting and reporting requirements to help countries, cities, and companies track how much waste is generated and where it ends up.[64]

"This standard is a real breakthrough," says Andrew Steer, president and CEO of the World Resources Institute. "For the first time, armed with the standard, countries and companies will be able to quantify how much food is lost and wasted, where it occurs, and report on it in ghly credible and consistent manner. There's simply no reason that so much food should be lost and wasted. Now we have a powerful new tool that will help governments and businesses save money, protect resources, and ensure more people get the food they need."[65]

Not only can this standard help governments and businesses track and report food loss and waste and develop strategies to reduce it, it can also bring them closer to meeting international commitments such as the Paris Agreement on climate change and the UN Sustainable Development Goals, which also call for a 50 percent global reduction in food waste by 2030.[66]

The losses associated with food waste are not limited to the food itself. Economies lose money, and the environment loses resources. Worst of all is the loss of lives, because there are so many suffering on this planet without enough to eat.

Agricultural Systems: Sustainability Is More Important Than Ever

Agriculture is a complex activity, and its sustainability depends on many factors, including the actual production of food, the use of energy (in particular, of fossil fuels), soil loss and depletion, and the availability and use of water resources.

Sustainable agriculture can be defined, briefly, as food production that makes the best use of nature's goods and services while not damaging these assets. Sustainable methods reject the practices of industrial food production and promote the health of the environment, economic profitability for farmworkers, and social and economic equality. Considering the harmful environmental impacts of agriculture that were explored in this section, the urgent need to embrace sustainability is more clear than it has ever been.[67]

Sustainable agriculture models are designed to benefit ecosystems by protecting soil against erosion, optimizing the use of water, minimizing the use of artificial fertilizers and fossil fuels, limiting the use of herbicides, fungicides, and pesticides, and encouraging biodiversity to reinforce the resilience of ecosystems.[68]

Revolutionizing the Practices of the Past

A critical part of implementing sustainable methods is rethinking existing models. Some models, such as the Green Revolution—a post–World War II effort to increase crop yields in countries threatened with famine by introducing farming methods that relied on agrochemicals, artificial fertilizers, hybrid seeds, and monoculture systems—have resulted in high productivity. But there have also been unintended consequences. Green Revolution practices have led to the often irreversible depletion of natural resources, including soil erosion, surface water contamination, groundwater pollution, deforestation, and loss of biodiversity.[69]

Beginning to change existing practices is key to halting damage to Earth's limited resources. Chemical-intensive monoculture systems, which involve growing identical plants over large areas season after season, may result in high yields, but the lack of genetic diversity over the years makes the plants vulnerable to disease and depletes the soil of nutrients. These effects increase the risk of massive crop failures and demand even greater use of artificial fertilizers.[70]

For example, until 1965 most of the world's bananas were of the Gros Michel strain. Unfortunately, around this time the Gros Michels were basically wiped out by Panama disease, a fungal infection that began in Central America and spread like wildfire across the world, wiping out commercial banana plantations as it went. The industry was forced to switch to the Cavendish banana, an "inferior" strain believed to be immune to Panama disease.

Today, the Cavendish strain is by far the primary banana grown and exported all over the world. Once again, though, the industry has created a banana monoculture. Recently, a new strain of Panama disease called Tropical Race 4 turned up in Malaysia, where it began to spread. The fear is that t will eventually cause as much damage as the original Panama disease. Tropical Race 4 has now appeared in Southeast Asia, Australia, and, as of 2013, Africa, threatening food security, livelihoods, and global banana stocks. This illustrates the danger of monocultures, especially with a fruit like the banana that is so crucial to feeding populations worldwide.[71]

By diversifying their cropping systems, farmers can replenish nutrients in the soil even as they reduce the risk of disease destroying an entire crop. Introducing more diverse crops also reduces pests, lessening the need for chemical pesticides and synthetic fertilizers. Legumes, such as peas, beans, and alfalfa, as well as ancient grains such as millet are not only nutritious food crops, they also contribute substantial amounts of the critical nutrient nitrogen to the soil. This nitrogen remains in the soil much longer than the nitrogen added via synthetic fertilizers and reduces water pollution caused by the use of these chemicals.[72]

Cultivating trees on farmland has been found to contribute to higher yields without the additional use of environmentally harmful chemicals. Over the past few decades, Nigerian farmers have planted native *Faidherbia* trees across 5 million hectares of farmland. The *Faidherbia* trees not only maintain nitrogen levels in the soil, they also protect fields from wind and water erosion and contribute beneficial organic matter to the soil when their leaves drop. By implementing this agroforestry system, the farmers have been able to double the maize yields of conventional plots. The success of this practice has caught the attention of farmers in Ethiopia, Kenya, and Zambia.[73]

In a report titled "The Future of Food and Agriculture: Trends and Challenges," the FAO explains that "high-input, resource-intensive farming systems, which have caused massive deforestation, water scarcities, soil depletion, and high levels of greenhouse gas emissions, cannot deliver sustainable food and agricultural production." FAO experts state that a future of monocultures must give way to "major transformations in agricultural systems, rural economies, and natural resource management, [which] will be needed if we are to meet the multiple challenges before us and realize the full potential of food and agriculture to ensure a secure and healthy future for all people and the entire planet."[74]

Farming for the Future

The increasing negative effects of climate change are proving detrimental to farming, creating massive obstacles to the maintenance of current methods and practices. Farming is therefore in the midst of a transitional period in which sustainable agricultural practices and methods are beginning to spread across the globe. The farming of the future aims to increase yields, reduce food scarcity, and decrease the detrimental environmental impacts of farming methods of the past.

"Emphasis will be on climate-smart agriculture in the short term, but in 10 to 20 years' time, the focus will be on switching crops," says Jason Clay, senior vice president of market transformation for the World Wildlife Fund. Alternative crops will be needed as climate affects current commercial crops. Clay cites sorghum as an example of a crop that is already being substituted for corn and maize, because it has multiple uses, including as an animal feed and even to produce beer. In Mexico, the government is attempting to use certain varieties of cocoa to replace coffee crops, which by 2025 may not be suitable to grow because of blight and the high temperatures caused by climate change.[75]

A huge instigator for change can be better investments in sustainable practices, such as crop rotation and adaptation. "Moving to adapted crop varieties that are more resilient to climate change is feasible," says Chris Brown, general manager for environmental sustainability at agribusiness Olam International. "But for the next wave of research breakthroughs, the FAO has estimated that we will need US$45–$50 billion in annual spending globally. It's currently at US$4 billion."[76]

Although switching crops could give smaller farmers who are struggling with their current crops the chance to outperform their previous efforts and become more productive, the right technical assistance is needed, along with packages of better genetics, management practices, and inputs.[77]

Enabling farmers to compete in the market and profit from sustainable methods requires training in technological innovation and adapting quickly to products that improve efficiency. Specialization allows farmers to focus their energy on niche products in order to survive in the face of competition from big name brands.[78]

The vertical farming practiced by the organization Sky Greens in Singapore has reduced the amount of land, energy, and water needed to grow crops. Vertical growers can produce 5 to 10 times more produce per unit area than traditional farms, and their crops can be grown year-round. This cuts down on cost and environmental impact and, considering that Singapore imports much of its food from China, allows the country to be more agriculturally self-sufficient.[79]

In addition to training and specialization, education is key to farmers recognizing and using the competitive advantages and disadvantages in their sustainable farming efforts. Policymakers are urged to promote education and training in sustainable farming, and new farmers are encouraged to establish sustainable technologies and invest long term in the relationship between profitability and sustainability. Universities and research centers are working with small family farms to find solutions to sustainability challenges and develop innovations.

Stephan Goetz, a professor of agriculture and regional economics at Penn State's College of Agricultural Sciences, discovered that "in certain regions of the country, community-focused agriculture has had a measurable effect on economic growth." His work will benefit policymakers and farmers looking to promote sustainable agriculture in local communities. Goetz found that "for every US$1 increase in agricultural sales, personal income rose by 22 cents over the course of five years," and thus he established that total agricultural sales and income growth are correlated.[80]

An additional report by Charles Benbrook, a Washington State University researcher, discovered through experimentation that organic foods have more antioxidants and less pesticide residue than conventionally grown crops, which benefits the health of the consumer tremendously in the long run. It was discovered that organic plants produce greater amounts of antioxidants and that plants that are not sprayed with pesticides produce more phenols, which help protect them from insects and disease and also help humans ward off infection when consumed. Benbrook explains the benefits of this study for the future of organic farming: "This study is telling a powerful story of how organic, plant-based foods are nutritionally superior and deliver bona fide health benefits."[81]

The world's farmers truly are the stewards of the land. The sustainability of the agri-food chain depends as much on the commitment of farmers, businesses, and policymakers as it does on the responsible choices of eaters. If all links in the food chain work together, they can have a powerful effect on the entire socioeconomic landscape of food production and consumption.

Conclusion and Action Plan

The importance of the food and agriculture sector is twofold. Although some food production methods are at the root of environmental and health challenges, the food and agriculture sector holds the greatest opportunity to solve a wide range of these problems, such as obesity, hunger, and climate change, for the current population and for the future.

The Double Pyramid illustrates the relationship between food choices and the environment. Based on the Mediterranean diet, this graphic learning tool shows that foods healthful for eaters, such as vegetables, are also healthful for the planet because of their low ecological footprint. Conversely, foods such as beef, which is high in cholesterol, is also one of the biggest contributors to GHG emissions.

The ecological footprints of various foods and their production, seen in the Double Pyramid, provide more in-depth information on the effects of food choices. The carbon footprint measures the effects on the climate of the amount of GHGs emitted through certain processes. The water footprint is similar, measuring the amount of water used and polluted all along the food production chain. The water footprint also brings the virtual water trade and the risks of water privatization to light.

Like air and water, Earth's soil quality and quantity have been depleted by irresponsible agri-food systems. The negative impacts on the environment have brought to the attention of farmers and policymakers the need to revolutionize the practices of the past.

Implementation of sustainable methods such as crop rotation and crop diversity has shown that they can reverse some of the damage caused by past methods. Better education and innovation in sustainability are promising for future farmers as they change their methods with hopes for a healthier environment, a healthier society, and greater economic profitability and equality.

With the right tools and better practices, a successful sustainable food system is achievable. The actions taken today—by eaters, farmers, policymakers, and investors—can have long-lasting effects on society and the planet. With hard work and commitment, the effects can be positive, ensuring a better future for everyone.

To help build a food system that supports a healthy planet

- Eaters can use the Double Pyramid to make food choices that are more healthful and more environmentally friendly.
- Water scarcity can be addressed by developing more democratic policies on water access and by adopting practices that will increase the productive and efficient use of water.

- Producers can combat climate change by switching to sustainable sources of energy. Eaters can make big differences through small changes to their diets.
- Farmers can reengineer their agricultural systems by allowing soil to regenerate and by reducing their reliance on environmentally damaging synthetic fertilizers. Investments must be made to train and educate farmers about more sustainable practices.

In December 2016, the BCFN and the Economist Intelligence Unit launched the Food Sustainability Index (FSI) at the 7th International Forum on Food and Nutrition in Milan, Italy. This newly developed ranking system is meant to create awareness, educate the public, and influence government leaders to make food and its production high-priority issues on their agendas. The FSI ranks countries' food sustainability systems based on three pillars: food loss and waste, sustainable agriculture, and nutritional challenges.

This one-of-a-kind index has the potential to revolutionize the way consumers, farmers, businesses, and ambassadors look at food and agriculture. Lucy Hurst, director of the Economist Intelligence Unit, claims,

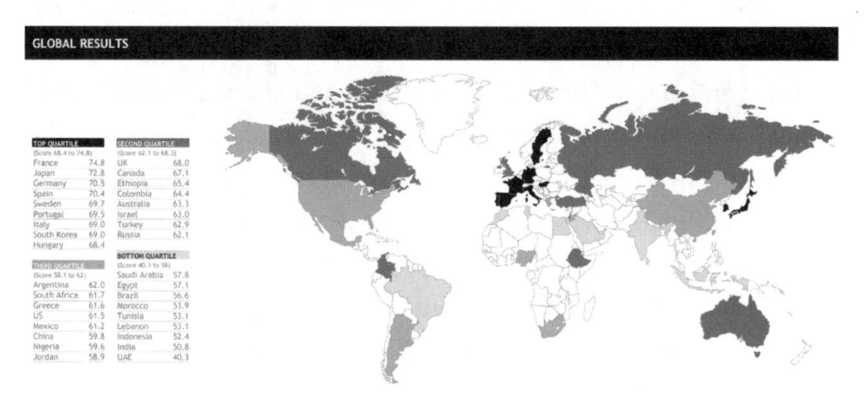

Figure 2.7. Measuring sustainability in the food system. Source: Food Sustainability Index, Economist Intelligence Unit and BCFN Foundation, http://foodsustainability.eiu.com/wp-content/uploads/sites/34/2017/11/Infographic-KEY-FINDINGS.pdf

"This index gives us the opportunity to look at three areas of food systems that we can all relate to—the quality of the food available to us, what we can do about food waste, and how critical sustainable agriculture is to providing enough food for us all. Learning from top performers can help countries meet their Sustainable Development Goals targets."[82]

The FSI's mission is also driven by some discouraging statistics. World population is projected to reach 8.1 billion by 2025, and 95 percent of the growth will come in developing countries, many of which are dealing with the double burden of hunger and rising obesity. Meanwhile, climate change is presenting new challenges to the agriculture sector. The objectives of the FSI, say researchers Francesca Allievi, Marta Antonelli, and Katarzyna Dembska, are "not only to highlight the performance of countries but to establish a comparable benchmark, to offer examples of best practices at the national and city levels, and to measure progress over time."[83]

The FSI analyzed agriculture, nutrition, and food waste in 25 countries, which together account for 87 percent of global GDP and 72 percent of the world's population, by using 58 different measures of sustainability. France, Japan, and Canada achieved the best scores for the production, distribution, and consumption of food. Their agriculture is the most sustainable, their food waste is lowest, and their diets are the most balanced and mindful of both people's health and the planet's well-being. France ranks above Japan and Canada in part because of its innovative policies to fight food waste and how it has encouraged a balanced diet in its population. For example, French supermarkets are required to donate excess food, and tax incentives are used as a way to discourage unhealthful food consumption. Japan and Canada rank second and third by virtue of their policies regarding sustainable agriculture and their widespread adoption of healthful, balanced diets.[84]

France, followed by Japan and South Korea, is at the top again for nutrition quality, largely because of its policy response to dietary patterns, such as instituting a tax on sugar in drinks. In 2012, France began taxing soda and other drinks with high sugar content, including sports drinks and juices. And in 2017, the country banned restaurants from offering unlimited soda drinks to customers. Companies have complied. Consider the fast-food burger chain Five Guys, which installed microchips in customers' cups so that if they go to the drink fountain more than once, it will automatically shut off.

The FSI identified Germany, Canada, and Japan as the countries with the best development and promotion of sustainable agriculture. Germany ranks first, with excellent results in how it manages water resources and its low use of pesticides and fertilizers. Canada comes in second thanks to the high scores it achieved in the quality of its subsidies, the diversification of agricultural production, and high productivity levels.[85] The evaluations made by the FSI also provide insight into the challenges faced by nations, both rich and poor, around the world.

The countries with the lowest overall scores are India, Saudi Arabia, and Egypt, with India and Egypt facing the double challenge of obesity and malnutrition. Their use of resources (especially water) is also considered unsustainable, and they are wasting and losing enormous amounts of food before it ever reaches consumers. According to estimates from both the Egyptian government and private institutions, anywhere from 20 to 50 percent of the country's wheat harvest is lost each year. Noting that poor infrastructure, inadequate storage, pests, and disease lead to these losses, the government began to take action. In 2015, it started a nationwide network of shounas, or grain storage facilities, to collect and store wheat. The shounas are temperature-controlled and will also serve as processing stations, allowing producers to clean, dry, and package wheat and other grains for sale. Since 2015, the shouna storage system has reduced grain waste to less than 5 percent, and Egypt has increased security measures within the storage system so that suppliers cannot blend in foreign wheat at Egypt's expense. The precedent that Egypt has set will soon be matched by storage security systems throughout much of the Middle East (Oman,

Saudi Arabia, Yemen, Algeria, and Tunisia) in order to cut back on grain waste. Egypt will additionally construct an export plant to produce parts for Middle East storage systems, which will employ 1,000 Egyptians.[86]

India, like Egypt, is in last place in part because of its unsustainable management of water resources and the inadequacies in the Indian diet. The country has the highest percentage of malnutrition among children younger than 5, and the World Health Organization estimates that half of all cases of malnutrition in India are caused by or associated with chronic diarrhea or chronic stomach infections from unclean water and a lack of proper sanitation.[87]

Unlike India, Saudi Arabia and Egypt are 24th and 23rd in the ranking, respectively, largely because of their excessive food waste and high levels of obesity. The United States and the Persian Gulf nations also have the highest per capita rates of food waste and obesity.[88]

The realities represented by these data are much more complex than the statistical numbers. According to researchers Allievi, Antonelli, and Dembska, "A food system does not sit in isolation, and a large number of stakeholders act together according to dynamics created by specific drivers. These include biophysical elements and constraints, innovation and research, political and economic inputs, socio-cultural aspects, and demographic issues. When scaling this picture to the regional, national, continental, and global level, it becomes increasingly complex, creating a high level of uncertainty when trying to assess the interaction among its parts."[89]

Guido Barilla, BCFN president, sees the FSI as an invaluable tool for better understanding the complexities of food systems. "The Food Sustainability Index will help us to understand where people eat the best around the world," he says, "not in terms of how good something tastes but in terms of the sustainability of the food system, helping researchers and decision makers to understand where to focus research and policy choices."[90,91] A process is now under way to apply the same measurements to the world's largest cities.

The Economist Intelligence Unit, in collaboration with the BCFN Foundation, has launched a pilot project, called City Monitor, to look at urban food systems. Through the evaluation of data and consumption habits, this new index is intended to identify indicators with which to better understand the dynamics of urban food systems. In the project's initial stage, it will look at 16 cities selected based on their geographic location, the availability of data, and their efforts to implement a sustainable food policy. The cities are London, Milan, Paris, Toronto, Belo Horizonte (Brazil), Johannesburg, Shanghai, Kyoto, Mexico City, Berlin, Moscow, Tel Aviv, Dubai, San Francisco, Lagos, and Mumbai.

According to Marta Antonelli, who helps lead the project at the BCFN, the center wants the FSI to catalyze effective change in urban food systems. Says Antonelli, "We mustn't forget that a number of studies have shown that interventions targeted directly at cities are more effective than those on a broader, national level. They're more focused and they have more attainable goals which are closer to the population." Antonelli thinks this kind of data is necessary for urban policymakers to make the best decisions possible regarding food access, affordability, and waste management on a city-specific basis. She explains, "Each city [will] adapt the sustainability projects to suit their own needs and priorities, but ultimately their experiences and achievements are shared within a global network—which can truly make a difference."[92]

Interestingly, City Monitor and the FSI are living indices, growing and expanding as leaders, farmers, and communities take action to solve food and agricultural challenges. And ultimately, they provide leaders with better tools for making more accurate measurements on all food-related issues, from nutrition to production and waste, and they also hold every party accountable for their actions. Countries with poor scores will have better information on how to make improvements, whereas the countries highly ranked on particular issues can serve as role models not only for nations and cities but for villages, towns, and individuals across the globe.

VOICES FROM THE NEW FOOD MOVEMENT: Dario Piselli

In what ways are young people instrumental in finding and utilizing sustainable solutions for agriculture and development?

It is often said that cities will be the greatest battleground for sustainable development in the 21st century, due in part to the continued increase in the world's population living in urban areas (projected to reach 66 percent by 2050). This is certainly true, yet it is crucial, for both ecosystem resilience and economic development, to ensure that we get the issue of sustainable agricultural systems right in the next few decades. Thriving rural areas, both in developed and developing countries, hold the key to solving the challenges lying at the interface of food security, land use change, energy production, biodiversity loss, air and water pollution, and so forth.

Unfortunately, this will not happen if the majority of young men and women living in these (often economically stagnant) regions are left without meaningful opportunities for skill development, access to resources and financial services, and decent employment. Rather, migration to urban areas or other countries will continue unchecked, with implications including growing urban poverty, social unrest, human trafficking, and conflict. So, in a fundamental sense, young people are instrumental in finding solutions for sustainable agriculture and local development simply because they are the most important stakeholders in this era of unprecedented changes in the world's economy. In both developed and developing countries, the global population now under 30, roughly half of the total, would suffer the direst consequences of a business-as-usual scenario for the agrifood sector.

Young people are instrumental in another sense of the word, as well. On the one hand, students and young professionals who have had access to education, finance, and proper support structures are now building profitable careers in all segments of agricultural value chains and creating robust livelihoods for themselves and their communities. On the other hand, even when they have not been exposed to formal education and business development services, rural youth in developing countries display a remarkable attitude towards innovation in agricultural practices. For example, there is increased use of information communication technology, swift uptake of climate-smart agriculture techniques, and sustainable use of resources in constrained environments.

The reason young people are instrumental, I would argue, is the same as when talking about the emergence of social entrepreneurship and the so-called sharing economy. Young people today, regardless of their country of origin, have seen their lives shaped by the consequences of short-term economic decisions and witnessed the negative impacts of environmental degradation and unsustainable development in their communities firsthand. This is why they exhibit such a strong sense of purpose and a natural attitude towards linking their individual role in the economy with the idea of working towards a broader social good.

Farmers and traditional farming techniques around the globe are aging. How can we make agriculture and the food system "cool" and more attractive for young people?

I think there are two ways of answering this question, depending on which perspective we adopt. In many developed countries, agriculture and the food system are already becoming "cool," be it because of the sense of purpose I was discussing earlier, because of new trends in consumer awareness, or because of the alternative that quality agriculture provides as an economic sector in times of severe youth unemployment. Make no mistake—we still have room for big improvements, and we must do a better job at increasing awareness about certain environmental impacts of food production that still go unnoticed, particularly in terms of overfishing and long-distance food transport. At the same time, the main challenge is now making agriculture more profitable by facilitating access to land, financial support, credit, and sufficient skilled labor for young farmers. In addition, we must promote adequate investments in non-farm careers that can create the demand and incentive for qualified young people to join the agrifood sector, including scientists, agronomists, and other non-farm entrepreneurs. Finally, we must respond to the problem of depopulating rural areas, which is a circular problem in that these areas become less attractive as more young people flee to the cities. Lack of planning, caused by increasingly urban-centric policymaking, can also be seen as a primary cause of the aging issue for farming.

When we look at developing and least-developed countries, making agriculture cool means ensuring that rural youth are not susceptible to the culture of consumerism, which created the problem in developed countries in the first place. However, the extent to which agriculture can be made cool in turn depends on the extent to which the agrifood sector can be made a hub for innovation, entrepreneurship, and opportunities. I doubt that poor working conditions, subsistence farming, and unskilled labor can be made cool anytime soon. Rather, we need to see decisive investments in access to knowledge and formal education, improved land tenure and facilitated land leases, adequate access to financial services and mentorship, and better connection with markets. Agriculture can be a business for a young entrepreneur in Africa, but governments have to create the conditions for it to happen. Engaging young farmers' associations in policy dialogues at the local and national levels can be the first step to ensure that these proposals are taken up. Another step is leveraging the role of international organizations, including the International Labor Organization and the UN Food and Agriculture Organization, in promoting policy coherence and capacity building within countries. Ultimately, however, it is the mobilization of domestic and multilateral finance, and the increase of private investments, that remain essential to generating business opportunities. From this perspective, the recent launch of the G20 Initiative on Rural Youth Employment is a very positive development, but it remains to be seen if its implementation will add value to existing efforts in this field.

Could you talk about your involvement as one of the founders of the youth division of the Sustainable Development Solutions Network (SDSN Youth) and discuss the major achievements since its launch in 2015?

My involvement with SDSN Youth began long before SDSN Youth was officially launched in June of 2015. I have been collaborating with SDSN since October of 2012, when I was a law student at the University of Siena focusing on international law and sustainable development. At the time, after expressing my interest in SDSN Mediterranean (Siena had just been appointed its host institution), I became part of the project team that was kick-starting its activities. I represented the youth constituency within its governance structure and worked to mobilize students on campus around the future Sustainable Development Goals (SDGs) framework. Then, in late 2014, Jeffrey Sachs expressed appreciation for the youth initiative of SDSN Australia/Pacific and asked his coordinator Siamak "Sam" Loni to turn that regional project into a global network. It was natural for Sam to assemble a core team of students and young professionals who were already involved in SDSN in different parts of the world to become the founding members of SDSN Youth as you know it today.

Regarding what we have achieved in little more than two years, I cannot deny it has been an impressive series of developments. From a small core team of around 10 volunteers who were already working full time in their careers or studies, SDSN Youth has grown to more than 100 staff in over 20 countries and a membership of over 300 member organizations in

65 countries. One of our main achievements is an ongoing project by my team on the Youth Solutions Report, which is a platform to support young innovators on the SDGs through better access to funding, expertise, and visibility.

The #KnowYourGoals campaign I coordinated in 2015, promoting the organization and adoption of the 2030 Agenda at over 120 events in more than 40 countries, has also been a major success. This is especially true given the fact that SDSN Youth had barely been launched at the time (actually, a second iteration of the campaign has just been announced). There are more, ranging from the hugely popular Vatican Youth Symposium held in October 2016 in partnership with the Pontifical Academy of Sciences, to the recent launch of the Local Pathways Fellowship, which seeks to empower young experts on urban sustainability to champion transformative change at the local level.

However, speaking in general, there are two overarching achievements that stand out. First, we are recognized as one of the leading and most respected organizations worldwide working on youth involvement in the 2030 Agenda. We are promoting our vision and goals in over 100 committees and working groups, including the UN Major Group on Children and Youth, the UN Policy and Strategy Group, the UN Working Group on Youth and Peacebuilding, the World Urban Campaign, and the Global Partnership on Sustainable Development Data. Second, we are doing this not just as advocates but as actual partners in the discussion and implementation of projects. In other words, we are furthering the view that there is more to youth involvement than advocacy and enthusiasm. There are skills and competencies, across youth communities around the world, which should be leveraged in support of sustainable development. Having the possibility of interacting with major organizations, governments, and companies from that perspective, I believe, truly changes the narrative about young people's place in solving the challenges of the 21st century.

Could you share a story that personally inspires you involving youth and their role in the Sustainable Development Goals?

Due to my position in SDSN Youth, I am exposed every day to inspiring stories of young people challenging all odds to work for sustainable development in their communities and regions. This includes fighting against environmental degradation, promoting gender equality, and championing new technologies. On a personal level, some of these stories hit closer to home than others. Through social media, I had the opportunity to learn about the incredible efforts of an Indonesian young conservationist and photographer, Pungky Nanda Pratama, who is now working in Sumatra as the education coordinator for an NGO called Animals Indonesia.

What strikes me the most about Pungky's work is the deep sense of wonder he puts into every photo of endangered species he and his team come across in the Sumatran forest. Also, the way he describes his attempts to educate the local population, especially kids, about the importance of ecosystem conservation. A few weeks ago, I happened to read one of his updates concerning changes in the attitude of local villagers towards wildlife trafficking. When a slow loris was found in the village, the village children immediately released it back into the wild. One might think this is a small story, but it shows how the dedication of many young people and environmental defenders, often persecuted and unknown, can really affect positive societal change towards wider sustainable development objectives. In a way, it reminds me of biologist E.O. Wilson's idea that being in awe of nature is a requisite condition for becoming passionate about its diversity, embarking on its study, and ultimately working to preserve its health.

What do you think young people can contribute to the Sustainable Development Goals in ways older adults cannot?

We are in an era of unprecedented transformation—in how our economies work, how societies cope with massive technological and other societal changes, and how our ecosystems are adapting and reacting to increased anthropogenic pressures. However, we seem to have forgotten that it is young people who are best positioned to analyze and solve these sorts of novel challenges. This is thanks to what scientists call fluid intelligence, a factor of general intelligence that was first described by Raymond Cattell and that usually peaks in young adulthood. Most of the greatest advancements in human history have arguably been the product of people in their twenties, so the idea that today the world is somehow too complex and should be left for older adults to sort out just does not add up.

It is easy to forget that young people today are the best-educated generation ever. Thanks to improvements in health and nutrition, youth are more intelligent than the average adult and are far more knowledgeable about new technologies. In addition, younger generations have a grasp of uncertainty and complexity that other age groups often lack. On the one hand, this leads to a better understanding of the synergies and trade-offs involved in addressing the cross-sectoral challenges enshrined in the 2030 Agenda for Sustainable Development. On the other, it allows young people to think of institutional arrangements and innovations that confront the many forms of path dependency that exist in international organizations, governments, and businesses and usually lead to inefficient, inequitable, and unsustainable outcomes.

What is very interesting is that for the first time, young people from different countries and regions share the same objectives and grievances, linked to the negative impacts of globalization and poor governance, and are increasingly part of a common culture. This goes beyond the usual notion that all young people are idealistic, even though idealism itself is everything but a negative word in the context of the major challenges we are facing. Rather, it speaks to the incredible, untapped potential of 1.8 billion people who are currently between the ages of 15 and 30 and largely hold the same ideas about how to transform our societies and economies for the better. Failing to partner with them would arguably represent the biggest waste of human capital in the history of mankind.

Dario Piselli is a sustainable development professional, young leader, and researcher. He is the Project Leader at SDSN Youth, the official youth initiative of the SDSN. He is also a research officer and Ph.D. student at the Graduate Institute of Geneva.

Citation: "Cultivating the Next Generation of Food Leaders: An Interview with Dario Piselli," Food Tank (2017). *Interview conducted by Brian Frederick in July 2017.*

VOICES FROM THE NEW FOOD MOVEMENT: Steve Brescia

As the executive director of Groundswell International and one of its co-founders, can you share with our readers what initially got you involved? What inspired you?

What initially inspired me, years ago, was working with social justice movements in Central America. To create just, democratic, and healthy societies, we need to create a foundation of farming and food systems that are of, by, and for the people. Since then, I have been constantly moved by the power of rural people in countries in the Americas, Africa, and Asia to organize, grow more food while regenerating their land, and create positive changes in their societies from the ground up. Over the years, we've identified ways of supporting improved farming and community development that work well and others that don't. We've seen that working to extend the technologies and logic of industrialized agriculture too often has negative impacts. Meanwhile, we've seen the effectiveness of farmer experimentation with ecological farming approaches, farmer-to-farmer sharing of locally generated solutions, and strengthening of community-based organizations lead to consistently positive outcomes. Many agricultural development practitioners', researchers', and farmers' movements are recognizing the power of farmer-led agroecology to allow people to improve their own lives. Working together, we have a real chance to address fundamental global problems like poverty, hunger, and climate change by unleashing and supporting the power of family farmers working productively with nature.

Groundswell International currently works in Burkina Faso, Ecuador, Ghana, Guatemala, Haiti, Honduras, Mali, and Nepal. Why were these countries selected as locations for sustainable farming development?

Groundswell International's co-founders represented local organizations from these countries. For decades, many of us have collaborated on people-centered development with rural communities in incredibly challenging circumstances. As these rural communities create better lives and improved farming production for themselves, they are also demonstrating the wider effectiveness of farmer-led agroecology. It is one reason we choose to work in very diverse ecological, cultural, and political contexts so that local solutions can strengthen national and global movements for sustainable farming and food systems. We are gradually growing to include other partners who are seeking to collaborate and build on these same principles.

How do Groundswell International's partnerships with nongovernmental organizations (NGOs) help to further a global farming mission?
If real and lasting change is going to occur, it has to be led by local people. Farming and food solutions need to be constructed by people where they live, in each context around the world, using both traditional and new knowledge. We can't just export the broken logic and technologies of our industrialized farming and food systems. So we very intentionally created Groundswell International as a partnership of local organizations and leaders. We tap into and build upon their wisdom and expertise and create programs together to promote positive and lasting changes. As partners, we all learn from each other and work to strengthen our abilities to create more sustainable farming and food systems in each country.

Food access and sustainability involve a wide variety of issues. What are some of the focus areas for Groundswell International projects?
We focus on strengthening and spreading agroecological farming and sustainable local food systems from the ground up. That means supporting people to experiment on their own farms to generate solutions and to spread both the ongoing learning process and successful practices

farmer-to-farmer. It means strengthening local organizations to manage this agricultural improvement, as well as marketing, savings and credit, and related work. It means strengthening the leadership of women. It increasingly means strengthening urban and rural relationships to promote local food economies. It also means connecting community-level initiatives to wider networks and social movements that are working to create policies that enable, rather than undermine, sustainable farming and local economies.

Groundswell International works to support sustainable agriculture around the world. How do you tailor your efforts to meet the food culture needs of each country that you serve?

When we start with and strengthen work led by local people, they take the lead in doing the work and generating the solutions themselves. People have a better understanding of their own needs and their own culture than we do, and they can and must have agency if positive and lasting changes are to be realized. We promote agroecological principles that allow farmers to innovate and work in productive, regenerative ways with nature, soil, seeds, water, and biodiversity. The result is more empowered communities creating healthier local food systems rooted in their culture and traditions.

Gender equity is emphasized through Groundswell International's mission and efforts. What unique role can women farmers play in sustainable farming?

In most of the places where we work, women are now doing the majority of the farming labor while also raising children and maintaining their families. While in many countries more men are being forced to migrate for work, women often remain more deeply connected to place, land, home, and family. Yet women are often excluded from the learning opportunities, resources, and decision-making power they need to improve their lives and those of their families. So any work to promote agroecological farming has to ensure women's involvement and leadership. As they nurture the land, they nurture their children. We recognize that we need to make certain that agroecological farming also results in improved nutrition for families, and women are key to that process. I've also seen how the successful farming practices of women can influence men—for example, women in West Africa convincing their husbands to adopt practices that regenerate land, rather than continuously clearing new land to farm. Supporting the leadership of women is essential to allow them to realize their own potential and to spreading sustainable farming and improve the well-being of families. The more we can do to support that, the more we can accomplish as an organization to bring about lasting change.

Is there anything you've learned from your experience with developing communities that could help developed countries, such as the United States, become more sustainable?

Absolutely. We are all human beings living on the same planet and face similar challenges. Of course, contexts vary widely, but effective principles apply across borders. We believe that as we work with farmers in places like Haiti or Burkina Faso, we are working with pioneers on the agriculture and food frontier. They are generating viable alternatives to improve their own lives under incredibly challenging circumstances. These experiences can be relevant to people in other contexts. How do we manage climate change and drought? How do we build resiliency in the face of shocks? How do we better connect farmers and consumers to provide healthy food? How do we regenerate soils and strengthen diverse local seed systems? There is a growing recognition that we need to transition from our broken food and farming system in the U.S. and the developed world to one that is more regenerative and in balance with nature, is healthier for people, and is more localized. That's a task we all share. We have much to learn from people in different contexts as we work to create this transition.

How can readers get more involved with Groundswell International?
Please reach out to us and let us know how you want to get involved. We need people with passion, creative ideas, energy, and resources who want to help build this community and create healthy farming and food systems together. We hope to hear from you.

Steve Brescia is the CEO and co-founder of Groundswell International, a group that promotes sustainable agriculture worldwide. Serving communities in Asia, Africa, and Latin America, Groundswell International provides resources for local farm development.

Citation: "Strengthening Farming Communities from the Ground Up: An Interview with Groundswell International," Food Tank (2015), https://foodtank.com/news/2015/09/strengthening-farming-communities-from-the-ground-up-an-interview-with-grou/. *Interview conducted by Brianna Marshall in September 2015 and edited by Michael Peñuelas in August 2017.*

VOICES FROM THE NEW FOOD MOVEMENT: Shaneica Lester and Anne-Teresa Birthwright

When did you know you wanted to become involved in agricultural research?

We have always been involved in agricultural research throughout our postgraduate journey. We worked on several local and regional projects centered on climate change and food security issues. However, we thought that there was a need for more action-oriented research targeting water challenges faced by small-scale farmers.

Where do you normally look to for ideas?

We identify the gaps, bottlenecks, and challenges in our society, particularly those which affect marginalized and vulnerable groups. Ideas emerge from there of how our knowledge and skills can be utilized in understanding and addressing these issues. As geographers, we also have a vested interest in finding innovative ways of improving sustainability for all.

Why did you decide to do participatory and action-based research?

We wanted to avoid a top-down approach and instead encourage self-empowerment within rural communities. A participatory approach allows farmers to be a part of their own solution by contributing their knowledge and expertise, as well as their perception and understanding of climate change. Action-oriented research allows us to focus on behaviors and practice, which are of major significance in building and sustaining local resilience.

What inspired the idea for this project?

Research has found that the Caribbean is expected to experience the impacts of climate change through increased variable rainfall, higher temperatures, and severe storm events, as well as an exacerbation of the warming and drying trends characteristically associated with El Niño events. For Jamaica, the agricultural sector is one of the many areas where climate change has increased unpredictability.

From farmers, we learned that climate change made it more difficult to efficiently irrigate crops, which affected farmers' productivity and willingness to continue in agriculture. It was also a deterrent to prospective farmers, especially youth. We thought that initiatives which involved the sourcing and implementation of various irrigation technologies might not be as far-reaching as building farmers' adaptation capacity through practice and self-perfection. Therefore, we believed integrating local traditional knowledge with technical and scientific know-how would be a more sustainable way of increasing adaptive capacity and productivity.

Why did you focus on coffee production?

Coffee is one of Jamaica's largest earners on the international market. The Blue Mountain and High Mountain (also called non–Blue Mountain) regions mark the island's two major grades of specialty coffee production. Blue Mountain coffee is produced at altitudes of 300 m (984 ft) to more than 1,500 m (4,921 ft) above sea level within the Blue Mountain range, while High Mountain coffee requires hilly areas above 300 m (984 ft) outside of the Blue Mountain range. Blue Mountain coffee holds a greater prestige and has the competitive advantage of being benchmarked by its taste, high quality, and appearance of bean; hence it's ranked the number-one brand on the international market.

Regardless of this notoriety, coffee farmers have not escaped the negative consequences of climate change. An International Trade Centre 2010 report stated that generally "coffee growers are by far the most numerous group that is directly affected and the most vulnerable" to climate change impacts. This is chiefly due to the significant influence that temperature and rainfall conditions have on potential coffee yield and crop development at various growth stages. The balance of these environmental factors is therefore integral in crop productivity and quality.

How have coffee farmers been specifically affected by climate change? Generally, most small-scale coffee farmers depend entirely on rainfall. Farmers now perceive, however, that there has been a decrease in rainfall and an increase in temperatures over the last 20 years. This has been noted especially during the April/May period where conditions at this time are crucial for the early flowering and berry development of coffee crops. Another perception among farmers is that the severity and frequency of droughts have increased. With water resources already stressed, farmers are therefore unable to effectively water their crops. This results in decreased coffee production and a delay in the bearing season.

Lately, farmers experienced two successive years of drought (2013/2014 and 2014/2015), which resulted in underdeveloped coffee beans. The variability of rains also increased the effects of the coffee leaf rust disease, amplifying the challenges faced by farmers to survive within the industry. With the island's main coffee variety—*Arabica typica*—being more susceptible to pests and diseases, as well as highly sensitive to increased temperatures, farmers are further disadvantaged.

Research has found that climate change may influence the locality and prevalence of certain pest and disease outbreaks due to changes in temperature and rainfall patterns. One such example is the coffee berry borer (*Hypothenemus hampei*). This is a small beetle that thrives only in coffee berries. Some farmers have reported an increase in the prevalence of the pest, as well as its occurrence at altitudes once absent, increasing their production costs.

Additionally, the prevalence of coffee leaf rust, caused by the fungal disease *Hemileia vastatrix*, has severely decreased production and threatened farmers' economic livelihood. In 2014, the International Coffee Council reported that the 2012/2013 crop year experienced one of the worst recorded outbreaks of leaf rust, with coffee production in some Central American countries (Costa Rica, Guatemala, Honduras, Nicaragua, and El Salvador) being severely affected. During the same crop year, the coffee industry in Jamaica also experienced its worst case of coffee leaf rust. According to farmers, the outbreak coincided with the passage of tropical storm Sandy in October 2012. It is believed that this was facilitated by the humid conditions which accompanied the storm, dispersing the fungal spores to areas and altitudes that previously had been unaffected.

What do you hope the outcomes of the research will be?

We hope that the farmers will start implementing the various methods they learn from the units of the irrigation knowledge transfer curriculum. We look forward to seeing more climate water conservation strategies, soil water management practice, and a greater understanding of plant–water interactions. From the effective use of these strategies, we expect that farmers will see an improvement in their crop health and yields. In successfully executing this project we hope to replicate the Knowledge Transfer Curriculum (KTC) in other agriculturally important communities.

Admittedly, in the midst of climate challenges, coffee farmers have remained resilient and taken action. It is precisely because farmers are vulnerable that they become autonomous, purposive actors who exercise adaptive strategies and negotiate their own resilience in the face of adverse circumstances. They therefore creatively empower themselves in the midst of unfavorable situations. Consequently, improving farmers' adaptive capacity through education, efficient infrastructure, financial safety nets, and social capital would reduce their vulnerability and enable farmers to not only survive shocks and stresses but become resilient and overcome future challenges.

Who has been helpful in making this idea a reality?
Moral support and encouragement from friends, family, and our mentors Dr. Robert Kinlocke (The University of the West Indies, Mona) and Dr. Kevon Rhiney (Rutgers, The State University of New Jersey) were instrumental for us pursuing this idea. But most importantly, the BCFN YES! research competition made this idea a reality by investing in our project. The Barilla Center sees the value young researchers can contribute to the sustainability and health of our food systems.

A year after your award, how have coffee farmers been adapting?
Jamaica's coffee farmers have implemented several coping/adaptive responses to the impacts of climate change. Some farmers are using the *Arabica geisha* (a more disease-tolerant and robust coffee cultivar) to combat the negative effects that the coffee leaf rust and rising temperatures have on the traditional *Arabica typica* variety. Because of the unpredictability of rainfall, farmers are also delaying fertilizing their crops, because fertilizer must be applied with water; otherwise, it will damage coffee plants.

Additionally, in an attempt to increase moisture retention and efficiently use the limited water available, farmers are applying a grass mulch to the roots of coffee plants and intercropping banana and plantain trees throughout the farm for added shade. Intercropping also allows farmers to generate additional income and contribute to household food security. Some farmers have also explored the decision to abandon their coffee farms for other valuable and exportable commodities such as sugarcane, cocoa, and yams, mainly because of their production throughout the year, inexpensive maintenance, and low initial financial investment.

Climate change will continue to push small-scale coffee farmers towards new realities, as they occupy a fragile space within an industry innately shaped by broader political, socioeconomic, and institutional forces. In an effort to address this vulnerability, many recent policy discussions have incorporated the idea of "building resilience" within rural communities. Building resilience may imply strategies to help farmers deal with shocks and stresses, or to "bounce back." Yet this concept has been broadened by some researchers to suggest that getting farmers back to their original system is not enough—the goal should be for farmers to come back stronger and better positioned for the future. It is therefore impossible to build resilience without first addressing—and transforming—the underlying power relations and social, economic, and political conditions which contribute to farmers' vulnerability in the first place.

What will your next steps be once your project is finished?
Upon completing the project we will also be at the end of our PhD journey. We look forward to becoming more involved in community development and issues of national interest. Additionally, we intend to explore how the irrigation knowledge transfer curriculum might be beneficial to other regions. We also look forward to working alongside the Barilla Foundation as alumni on issues of food, nutrition, and environment.

Since 2012, the BCFN Foundation has hosted Young Earth Solutions (YES!), an annual competition that inspires young researchers under 35 to conduct innovative studies to meet global research needs on the sustainability of food systems. In 2016, Jamaican researchers Shaneica Lester and Anne-Teresa Birthwright won the competition and a 1-year research grant for the amount of 20,000 (US$21,367). Their project focused on small farmers' exploration of various climate adaptive irrigation strategies.

Citation: "Voices from the New Food Movement: Shaneica Lester and Anne-Teresa Birthwright," Food Tank (2017). *Interview edited by Michael Peñuelas in July 2017.*

CHAPTER 3

Food for Health

GOOD HEALTH REQUIRES HEALTHY, NUTRIENT-RICH FOOD, and a nutritious diet requires a team effort. Parents, children, public health practitioners, food producers, policymakers, retailers, funders and donors, and research institutions all play important roles.

Although individual eaters have the power to make choices about their own health, those choices are often influenced and informed by outside forces, including Big Food lobbyists and agribusiness. Decisions made by policymakers, for-profit corporations, and advertising campaigns can shape how and what consumers eat. Too often, these forces work together to make healthful foods the less attractive option, particularly for young eaters.

While nutrition advocates continue to call for increased access to healthful and affordable foods, they are competing against a multi-trillion-dollar global food and beverage industry, which spends billions each year to shape consumer demand by marketing to children and tens of millions more lobbying for a favorable regulatory environment. A number of the industry's largest players have been caught engaging in duplicitous conduct, including funding public health groups while lobbying against the policy actions that those same groups recommend and paying expert academic researchers to downplay the health risks of products in peer-reviewed publications.[1,2,3,4,5]

Fortunately, good food advocates such as Sam Kass, an American chef, entrepreneur, and nutrition advocate, are fighting for better health.

In the early 2000s, Kass developed a strong relationship with Michelle and Barack Obama when he served as their personal chef in Chicago. Later, when the Obamas went to Washington, Kass went with them and became assistant chef at the White House. In 2009, with then First Lady Michelle Obama, he built the first large-scale vegetable garden at the White House—1,500 square feet of organically grown crops for both White House meals and local food banks—since Eleanor Roosevelt's Victory Garden. Kass was active beyond the kitchen, though, and as the senior policy advisor for nutrition to President Obama, he used his visibility to raise awareness about issues of food and nutrition. "You look around our country and you see that we have a lot of major challenges, the origin of which is food," he said in an interview. "It's not a big step to think about: What am I doing? How is that affecting this problem? How am I helping?"[6,7]

In 2010, Kass also became director of the First Lady's Let's Move! campaign, an initiative focused on raising awareness about the problems associated with childhood obesity and about promoting healthier, more active lives for children. His direction helped turn the office of the First Lady into a bridge between the food industry and health advocates. Let's Move! not only offered tips for healthier lifestyles, it also pushed for requiring calorie counts on menus and collaborated with the food and beverage industry to reduce sodium content in their products and include calorie totals on food labels.[8,9]

Kass's leadership contributed to passage of the Healthy, Hunger-Free Kids Act, which required the U.S. Department of Agriculture to set new nutritional standards for all food served in schools. It called for increasing the amount of fruits, vegetables, and whole grains in meals, decreasing

Table 3.1. Optimal average energy in children's diets: median kcal/day multiplied for physical activity level for respective age group.

Age of the child	Optimal Average Quantitative Amount of Energy
2–3	1,050–1,470
4–6	1,350–1,640
7–10	1,620–2,300

Source: LARN 2014, www.sinu.it/html/pag/02-Fabbisogno-energetico-medio-AR-nell-intervallo-d-eta-1-17-anni.asp

Table 3.2. Energy requirements during adolescence (median energy in kcal/day per PAL).

	Median energy in KCAL/Day per pal	
Age	Male	Female
11–12	2,440–2,600	2,210–2,340
13–14	2,780–2,960	2,440–2,490
15–16	3,110–3,210	2,510
17*	3,260	2,510

*Over 18 years, kilocalories are calculated according to height and weight. Source: LARN 2014, www.sinu.it/html/pag/02-Fabbisogno-energetico-medio-AR-nell-intervallo-d-eta-1-17-anni.asp

the amount of sodium and sugar, and banning trans fats. More broadly, it helped shine a spotlight on childhood obesity and overall health that is helping parents, school administrators, and students understand the health impacts of poor nutrition and encouraged greater public interest in healthful meals in schools.[10,11]

Efforts to improve childhood nutrition are taking place all over the globe, and schools are usually at the center of these efforts. In Dakar, Senegal, for example, the majority of residents depend on foods imported from other parts of the world. This dependency means not only that residents consume more packaged, processed foods but also that they are losing touch with local food varieties and culinary traditions.[12]

Seck Madieng has been a leader in reconnecting Senegal with its food roots. A journalist by training, he left a career as a government bureaucrat to pursue more meaningful goals that spoke to his interest in farmers and how they work and eat. In 2006, he started *Agri-Infos*, the only Senegalese newspaper to focus entirely on agriculture, food, and healthful diets.[13]

In 2007, Madieng, along with a local chef, Bineta Diallo, started the Mangeons Local ("Eat Locally") education program in two schools in central Dakar. Mangeons Local pursues its mission—teaching students about how foods are made and who grows and prepares them—through classroom lessons that focus on local foods and regional culinary traditions. The program also teaches students how to cook, many for the first time, according to Diallo. Instead of eating baguettes and imported canned foods, the children learn how to cook cereals and grains, including local rice varieties, fonio (a small grain typically used in couscous), millet, and sorghum. And rather than drinking milk out of boxes imported from the Netherlands, they learn how good local milk can taste and about all the things that can be made from dairy products, including cheese and butter.[14]

"In a country like ours, where a majority of people live in poverty and we import most of our staple foods," said Madieng, "using local products can help families save money, even a little at a time, which can add up in the long run." Madieng also thinks that children, by bringing home the skills they learn at school, are the best vehicle for communicating information about food and nutrition to parents and helping to improve their families' diets. Mangeons Local hosts an annual community festival at each school, raising more awareness among families and the broader public.[15,16]

Creating alliances and partnerships is why programs and projects such as Mangeons Local, Slow Food International, and Let's Move!

are important. It is often hard for individuals, families, and schools to change their eating habits alone, but campaigns that teach children and adults alike about the benefits of healthy eating can have a huge impact.

In this chapter we discuss how good nutrition and a healthy lifestyle are essential parts of a longer, healthier life. We will explore what makes up a healthy diet, the way our dietary needs change throughout our lives, and the health risks resulting from a poor diet, in the paradoxical form of both hunger and obesity. It will examine how the dietary choices made by eaters can be influenced by profit-driven groups, including Big Food and agribusiness. This section will also highlight the decisions and actions eaters and consumers can make to influence world health and help create a healthier food system.

Building the Foundation for Health

Decades of studies have brought the scientific community to this simple conclusion: There are lifestyles and ways of eating capable of reducing the health problems that negatively affect quality of life and mortality rates, from obesity and diabetes to cardiocirculatory diseases and cancers. When combined with nutritious food choices, a lifestyle that includes physical activity is most likely to result in a life free of disease and chronic health problems.

Choices made by individual eaters ultimately determine the quality of their health over their lifetimes, and the knowledge to make the right food choices can be more beneficial if it's learned at an early age. Lifestyles and behaviors acquired at an early, impressionable age, such as dietary preferences, mealtimes throughout the day, portion sizes, and the tendency toward an active or sedentary lifestyle, are important factors in creating overall dietary behavior for life.

A study conducted by Tanda Kidd and Paula Peters, associate professors of human nutrition at Kansas State University, found that the dietary habits and behaviors adopted during the first few years of life are decisive influences on a person's health in later years. "Nearly one in four children ages 2 to 5 is overweight or obese," said Peters. "An obese child is at risk for developing diabetes, high blood pressure, asthma, and sleep apnea. A primary key to teaching a child to make healthy food choices is to start early." Kidd added that parents should encourage opportunities for active play several times a day and that "kids 6 and over should be physically active at least 60 minutes a day" in order to maintain a healthy and balanced lifestyle.[17]

Sarah Couch, a professor in the Department of Nutritional Sciences at the University of Cincinnati Medical Center, has found that parents who encourage healthful eating practices, set food rules at home, and provide healthful food weigh less overall themselves. Her research has concluded that not only encouraging healthy eating but also modeling that behavior increases the likelihood that a child will maintain better eating habits in the future.[18]

As Claudio Maffeis, a pediatrician, pointed out at the Barilla Second International Forum on Food and Nutrition, "The earliest years of life are a very important window in terms of the development of the organism. . . . Eating right during the developing years is important because it not only ensures that the child will grow and develop properly, but it also guarantees a defense against diseases, metabolic and otherwise, that we might encounter in later phases."[19]

A child who develops healthful habits is more likely to have the same healthful habits, and fewer health problems, as an adult. A 2014 study by Kathryn Hesketh, an epidemiologist at the Clinical Health Collaboration Centre of Excellence for Diet and Activity Research in the United Kingdom, concluded that a 4-year-old's level of physical activity was directly correlated with how physically active that child's mother was.[20] So providing targeted educational opportunities to the mothers of young children can be key to increasing the overall health of both mother and child.

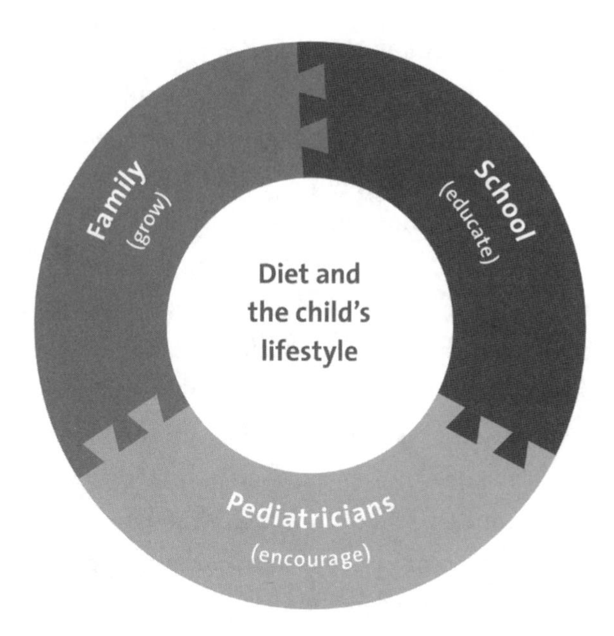

Figure 3.1. Actors in food education. Source: BCFN, 2010, from p. 181 of *Eating Planet*, 2nd ed. (Milan: Edizioni Ambiente and the Barilla Center for Food & Nutrition, 2016).

Targeted educational opportunities are promoted by organizations such as Bent on Learning and REAL School Gardens, which are working to improve the healthy habits of youths. Bent on Learning spearheads a campaign to bring yoga to schools in New York City, leading more than 2,000 kids per week through exercises in their classrooms. REAL School Gardens has built urban gardens next to more than 100 schools in Texas and the mid-Atlantic region of the United States, where it teaches kids to grow—and eat—their own organic produce.

Another organization, FoodCorps, places AmeriCorps service members in limited-resource school systems across the United States to work with teachers and students to establish farm-to-school programs, incorporate nutrition education into school curricula, and plant school gardens. A 2015–2016 evaluation of FoodCorps by Columbia University's Tisch Food Center at Teachers College found that 75 percent of participating schools were measurably healthier school food environments after 1 year in the program and that students in participating schools were eating triple the amount of fruits and vegetables they had been before.[21,22]

Aviva Must, a professor of public health and community medicine at Tufts University, said that instilling healthful eating habits in children and adolescents is a shared responsibility and that schools, in particular, carry much of that responsibility. "Schools are important reinforcing environments," she said. "In the U.S., schools no longer teach cooking skills as part of secondary school curricula; although it would take a lot to bring it back, it deserves serious consideration. The lack of cooking skills represents an important impediment to healthful eating at home, where increasingly, youth prepare their own meals."[23]

Educating children about the importance of a good diet can lead to future generations that are healthier, live longer, and have fewer illnesses. As studies and experts from around the globe have found, the foundation for a long and healthy life must be built at an early age. Young people are impressionable, so it is crucial that the responsibility of building this foundation is a concern of all, especially when looking toward a healthier future.

A Lifetime of Health, and Preventing the Paradox

Life is typically viewed as a series of phases: infancy, childhood, adolescence, adulthood, and, eventually, old age. A nutritious diet, along with physical activity, during the earlier phases of life is crucial to healthy development, but the benefits affect more than a person's physical growth. The habits learned early in life support health maintenance throughout adulthood and are the leading contributors to the prevention of diseases and ill health into old age.

Box 3.1. Tools for Promoting Well-Being

Gross domestic product (GDP) is a quantitative measure of macroeconomic activity. It reflects the volume of economic activity of a country and approximates the wealth and economic well-being of its citizens. Wealth is still used as a tool to measure well-being, when in fact many other elements should be included in assessing quality of life: social inclusion, equality, the state of the environment, health issues, and the lifestyles of individuals and societies.

The Barilla Center for Food & Nutrition (BCFN) has contributed to the subject of well-being as it relates to the center's principal areas of research and analysis: nutrition and its impact on quality of life. Consider first the effect of food choices on the health of children and adults, both negative (direct causes or risk factors for serious disease) and positive (protection against certain diseases). However, the impact of food and nutrition on our environment is also significant, because of the way food production consumes and degrades natural resources (through soil depletion, water use and pollution, and the emission of air pollutants, including GHGs).

The BCFN's work has resulted in three multidimensional summary indices for the quantitative measurement of national well-being. The first is the Index of Current Well-Being, which measures the present well-being of individuals, how people are feeling and living today. The second is the Index of Well-Being Sustainability, which measures the dynamics and future trends of the current level of well-being. And the final index is the Food Sustainability Index (FSI), which was produced in collaboration with The Economist Intelligence Unit. It has three pillars, including one focused on nutritional challenges faced by countries.

The Index of Current Well-Being and the Index of Well-Being Sustainability were used to measure the performance of each nation in seven dimensions: psychophysical and behavioral well-being, subjective well-being, material well-being, environmental well-being, educational well-being, social well-being, and political well-being.

In the Index of Current Well-Being, Sweden ranked highest, with 6.9 points (out of a possible 10), followed closely by Denmark, with 6.5 points. The United Kingdom came in third, with 6.1 points, followed by France (5.5) and then a trio of countries with similar scores: the United States (5.3), Japan (5.2), and Germany (5.1). Italy maintained the same position and point total

(4.9) it registered in the previous BCFN Index. Spain and Greece finished in ninth and tenth place, with 4.3 and 3.5 points, respectively.

The Index of Well-Being Sustainability was also an aggregate index, consisting of 25 performance indicators to measure the seven dimensions of well-being in three subindices: the lifestyle subindex, the wealth and environmental subindex, and the social and interpersonal subindex.

The Index of Well-Being Sustainability was led by Denmark, with 7.4 points, followed closely by Sweden, with 7.3 points; third place belonged to France (6.3) and fourth to Germany (6.1). Sitting between 5.9 and 5.3 points, in descending order, were Spain, Japan, the United Kingdom, and the United States, and Italy found itself in the penultimate spot, with 5.1 points (although the distance between Italy and Germany is only 1 full point). Greece once again came in last, with 4 points, separated from the other countries by a substantial gap.

Well-being has a large, multidimensional role in every life, and measuring it more accurately allows a better perception on the quality of life. The two indices developed by the BCFN are capable of doing what the GDP fails to do. In 1968, Robert Kennedy remarked on the function of the gross national product (GNP), a close cousin to GDP, as a measurement tool: "It measures everything, in short, except that which makes life worthwhile. And it can tell us everything about America except why we are proud that we are Americans."[a]

a. R.F. Kennedy, "Remarks at the University of Kansas, March 18, 1968," The John F. Kennedy Presidential Library and Museum (1968), https://www.jfklibrary.org/Research/Research-Aids/Ready-Reference/RFK-Speeches/Remarks-of-Robert-F-Kennedy-at-the-University-of-Kansas-March-18-1968.aspx

Infants, children, and adolescents without access to proper nutrition are especially vulnerable to a lifetime of poor health. Serious problems in the growth of a child, such as osteoporosis, iron deficiency, and high blood pressure, can result from a chronically poor diet.[24]

Around the world, undernutrition is the root cause of 45 percent of all child deaths. Developing countries are hit the hardest by this global problem. Southeast Asia and the Indian subcontinent have greater numbers of underweight children than any other part of the world. The

GUIDELINES FOR CARDIOVASCULAR PREVENTION

Fats: 15-30% of total calories	Saturated fats <10% and trans fats <1%	4-5 portions of fruit and vegetables daily	1-2 portions of fish every week	Encourage the consumption of unrefined cereal grains	Consumption of alcohol not recommended
30 minutes of physical activity every day	Less than 140 g of meat a day	4-5 portions of legumes a week	Avoid conditions of overweight and obesity	Don't smoke	Salt: 4-5 g/day and no dietary supplements

GUIDELINES FOR PREVENTION OF DIABETES

Fats: <30% of total calories	Saturated fats <10% and trans fats <1%	5 portions of fruit and vegetables daily	2-3 portions of fish every week	Encourage the consumption of unrefined cereal grains	Consumption of alcohol not recommended
150 minutes of physical activity every week	Proteins: 10-20% of total calories	4 portions of legumes a week	Avoid conditions of overweight and obesity	Maintain a normal BMI	Salt: 6 g/day and no dietary supplements

GUIDELINES FOR PREVENTION OF TUMORS

Limit consumption of fats	Don't smoke	5 portions of fruit and vegetables daily	Prefer fish to red meat	Encourage the consumption of unrefined cereal grains	No more than one glass of alcohol per day
45-60 minutes of physical activity every day	Limit consumption of red meat and salami	Eat legumes regularly	Avoid conditions of overweight and obesity	Maintain a normal BMI	Moderate salt intake

CONVERGENCE OF THE GUIDELINES
BARILLA CENTER FOR FOOD & NUTRITION

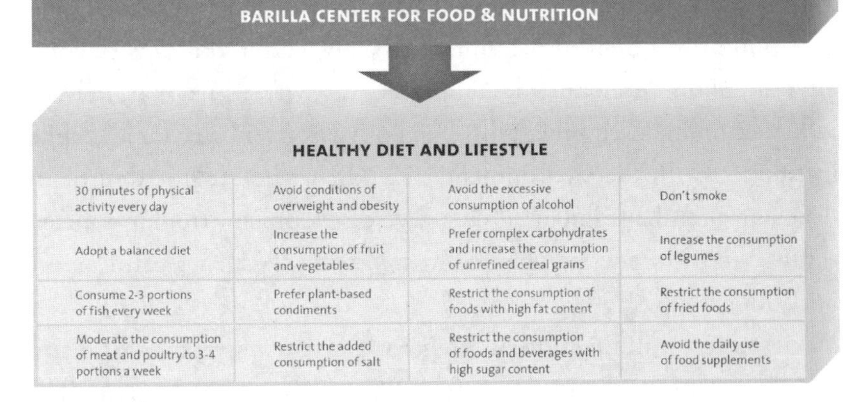

HEALTHY DIET AND LIFESTYLE

30 minutes of physical activity every day	Avoid conditions of overweight and obesity	Avoid the excessive consumption of alcohol	Don't smoke
Adopt a balanced diet	Increase the consumption of fruit and vegetables	Prefer complex carbohydrates and increase the consumption of unrefined cereal grains	Increase the consumption of legumes
Consume 2-3 portions of fish every week	Prefer plant-based condiments	Restrict the consumption of foods with high fat content	Restrict the consumption of fried foods
Moderate the consumption of meat and poultry to 3-4 portions a week	Restrict the added consumption of salt	Restrict the consumption of foods and beverages with high sugar content	Avoid the daily use of food supplements

Figure 3.2. Guidelines for food and nutrition. Source: BCFN, 2009, from p. 165 of *Eating Planet*, 2nd ed. (Milan: Edizioni Ambiente and the Barilla Center for Food & Nutrition, 2016).

United Nations has estimated that 1.2 million Indian children die each year before reaching 5 years of age—that's more than two children a minute—and the World Health Organization has estimated that about 30 percent of children under 5 in India are underweight. The Indian Ministry of Health and Family Welfare estimates that only 9.6 percent of children under 2 years of age receive an "adequate diet."[25,26,27,28]

Anthony Lake, executive director of the United Nations Children's Fund, said, "Every year, undernutrition contributes to the deaths of around 3 million children and threatens the futures of hundreds of millions more—undermining the healthy development of their bodies and their brains, and affecting their ability to learn and to earn later as adults. And undernutrition doesn't affect only the health and well-being of individual children. By preventing children from reaching their full potential, undernutrition also undermines the strength of their societies."[29]

Malnutrition affects not only children's bodies but also their minds. A study conducted at the Division of Nutritional Sciences at Cornell University found that "an estimated 200 million children under the age of 5 years in low- and middle-income countries are at risk of not achieving full developmental potential partly due to undernutrition. Malnutrition affects brain development directly, and also affects physical growth, motor development, and physical activity, which may, in turn, influence brain development through both caregiver behavior and child interaction with the environment." Children who are micronutrient-deficient often have impaired cognitive development, trouble concentrating, and low energy when compared with classmates eating more nutritious foods.[30,31]

Malnutrition in the form of hunger is a frightening reality for many, but others suffer malnutrition's effects in a different form: obesity. The increasingly sedentary lifestyles of modern eaters, who do not need or burn as much energy as their early ancestors did, is contributing to the rise of chronic diseases. In fact, today's children may experience a worse quality of life than their parents as a result of poor nutrition, noncommunicable diseases, and lifestyle changes.[32]

This construct of nature poses an interesting paradox: While much of the world is suffering from malnutrition because of lack of food, an equal number struggle with the effects of the overconsumption of foods that neither nourish nor contribute to healthy growth. Though vastly different circumstances, they are both enemies of a lifetime of good health.

Around the world, 41 million children under the age of 5 were considered unhealthily overweight or obese in 2014, with a body mass index (BMI) of at least 25. However, having enough food to eat can still result in malnourishment. *Micronutrient deficiency* is the term describing those who consume enough calories but are not getting the vitamins and minerals they need. They may have access to an adequate amount of food, but the nutritional quality is very poor. This food-related health issue is, again, most often seen in developing countries, including Mali, Nepal, Afghanistan, and the Democratic Republic of Congo.[33,34,35]

Margaret Chan, director general for the World Health Organization from 2007 to 2017, stated in a speech, "For the first time in history, rapidly growing prosperity is making many previously poor people sick. This is happening in countries with few resources and health system capacities to respond. If current trends continue, a costly disease like diabetes can devour the gains of economic development."[36]

The prosperity of developed countries does not guarantee a nutritious diet, though. In fact, three-quarters of American children do not consume enough vitamin D, two-thirds do not consume enough vitamin E, and more than a quarter lack the recommended amounts of calcium, magnesium, or vitamin A in their diets.[37]

A diet lacking in proper nutrition is detrimental to a child's growth. A 2014 study by Wilder Research connected good nutrients to better thinking skills, behavior, and overall health. Fifth-grade students were found to perform worse on standardized math and reading tests when their diets

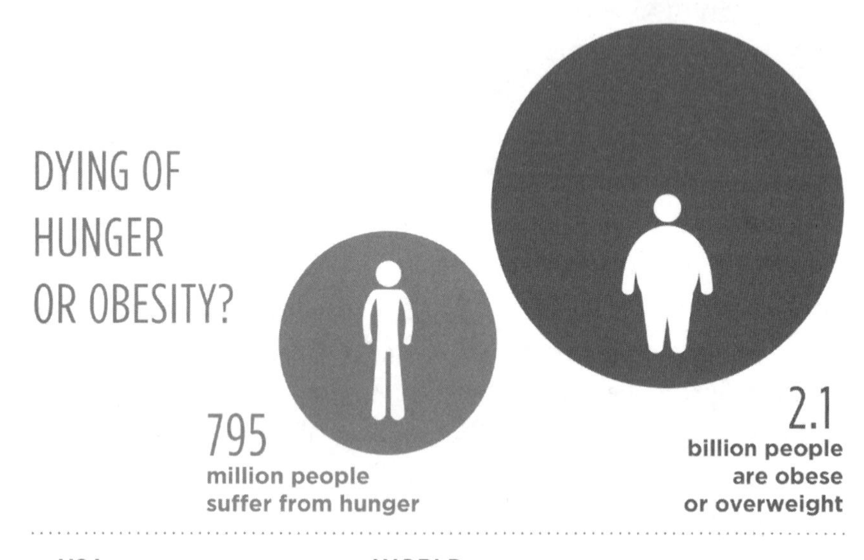

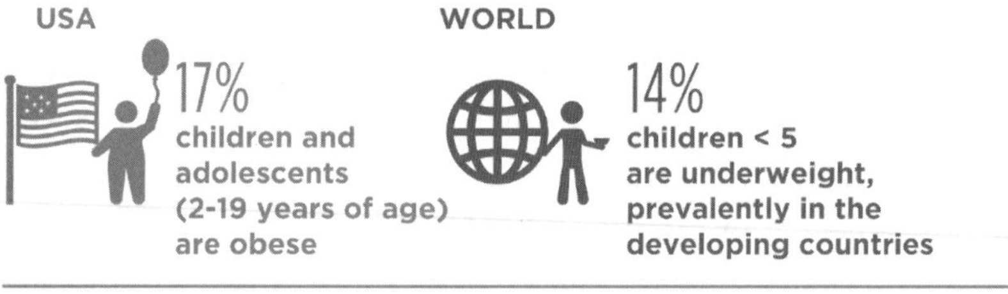

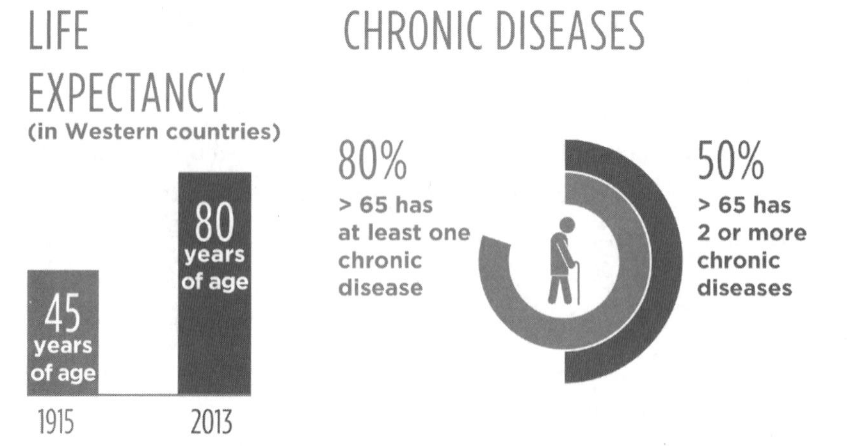

Figure 3.3. Food and health.

FORECAST OF DISEASE TYPE

in millions

ESTIMATE NEW DIAGNOSIS		ESTIMATE DEATHS
DIABETES	**CANCER**	**CARDIOVASCULAR**
592 (mln)	22 (mln)	23.6 (mln)
2035	2030	2030

MORTALITY IN THE WORLD

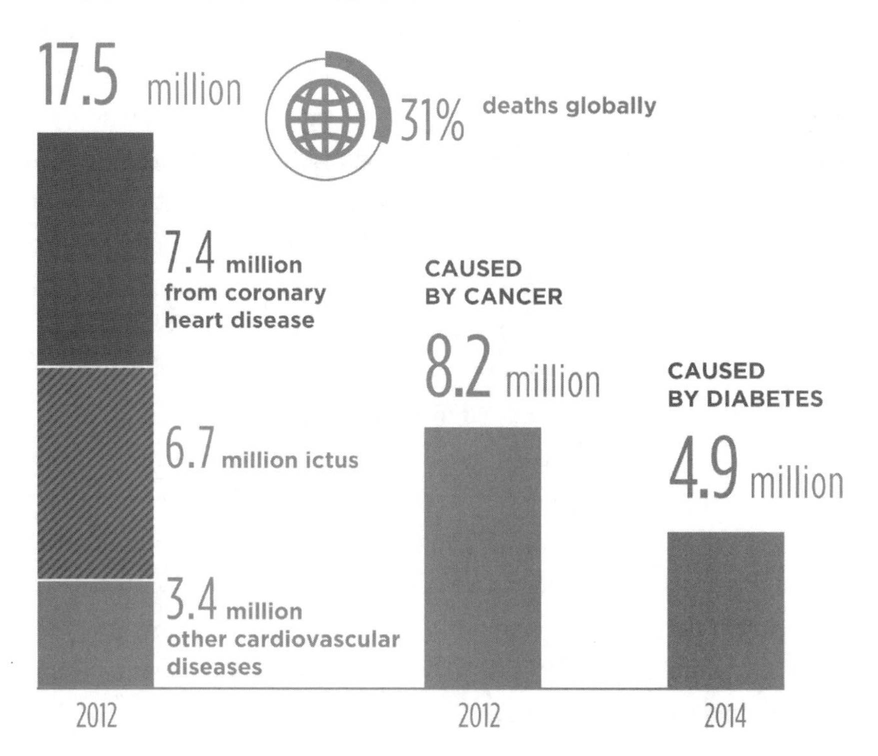

were less nutritious and heavy in fast food. Schools that have banned junk food have seen an increase in student scores on standardized English and science tests. The study concluded that children who eat a healthful diet, starting with breakfast before school and continuing throughout the course of the school day, have overall better cognitive functions.[38]

Dr. David Blumenthal, president and CEO of the Commonwealth Fund, a healthcare philanthropy in New York City, states that the greatest risk factors for children and adolescents are parental weight, intensity of the food marketing they are exposed to, and the types of food being sold where they live. Blumenthal advocates for organizations targeting this issue. "Programs that educate consumers generally, and children and adolescents in particular, and encourage all of us to engage in healthy lifestyle habits such as physical activity have the potential to pay large societal dividends, improving the health and well-being of our citizens and returning some of the savings our healthcare system desperately needs," he said.[39,40]

Fortunately, programs of this nature are growing in number. Action for Healthy Kids is working in 29,000 schools across the United States. Action for Healthy Kids helps school administrators audit the health of their students, assists in the creation of school-specific action plans, and provides grant money to implement health-focused curriculum interventions.[41]

Without intervention and proper nutritional education, children can carry the risks associated with both hunger and obesity into adulthood and, probably, for the remainder of their lives, creating a new generation of citizens at the mercy of the paradox, afflicted with noncommunicable diseases and cognitive disabilities. The effects of increasing obesity rates on today's adults, as well as consistently high hunger rates, underscore the importance of healthful, preventive eating and access to nutritious foods for all.

According to the UN Food and Agriculture Organization, an estimated 815 million of the 7.5 billion people on Earth were suffering from chronic undernourishment in 2017.[42]

Mexico, with the second-highest adult obesity rate in the world, affecting 32.4 percent of its population, provides a good example of this paradox. Meanwhile, 14 percent of Mexicans overall report having food access insecurities, with 10.8 percent having severe food access insecurities. The burden of food insecurity disproportionately affects the country's rural areas, where 33 percent of indigenous children are hungry. In 2010, the issue of food shortages was so drastic that almost 14 percent of Mexican children under 5 suffered stunted growth.[43,44,45]

Globally, however, there are now more obese people, and more obesity-related deaths, than there are underweight people and starvation-related deaths, excluding parts of sub-Saharan Africa and Asia.[46]

Research led by scientists at Imperial College London shows that the number of overweight adults in the world has grown to outnumber those who are underweight. Lead author Majid Ezzat, chair of the Global Environmental Health Department at Imperial College, stated, "Our research has shown that over 40 years, we have transitioned from a world in which underweight prevalence was more than double that of obesity to one in which more people are obese than underweight. . . . [G]lobal obesity has reached a crisis point. We hope these findings create an imperative to shift responsibility from the individual to governments and to develop and implement policies to address obesity. For instance, unless we make healthy food options like fresh fruits and vegetables affordable for everyone and increase the price of unhealthy processed foods, the situation is unlikely to change."[47,48]

Whereas 36 million people around the world die every year from malnutrition and famine, 3.4 million die from being overweight. An estimated 44 percent of them die of diabetes-related illnesses, 23 percent of heart attacks, and about 40 percent of tumors—all attributed to overeating or a poor diet.[49,50]

Interestingly, both life expectancy and the rate of chronic disease are simultaneously on the rise. Thanks to modern medicine, humankind may soon experience, for the first time in modern history, prolonged old age marred by fragility, disability, and poor health. It is estimated that by 2025, the old will outnumber the young, with the population of adults over 60 exceeding 1.2 billion.[51,52]

Living longer does not necessarily guarantee a healthy life in old age, and this creates more burdens on healthcare systems because of the higher healthcare costs for obese older adults. A study of obese Medicare beneficiaries estimated that in 2009, obese older adults cost Medicare approximately 8.5 percent more than older adults of average size, with the greatest cost going toward drugs for obesity-related conditions.[53]

Undernutrition, too, creates great risks for older people and illustrates how the paradox of hunger and obesity can affect eaters at any age. Older people naturally eat less, and with age their need for energy is reduced. This occurs simultaneously with physiological changes in their sense of taste and smell, which can cause them to eat either not enough food or a diet lacking in micronutrients. Older adults who are immobile or who have limited social interactions often adopt a nutrient-poor diet unintentionally. The consequences of undernutrition in older adults include vulnerability to infection, increased falls, loss of energy and mobility, and confusion.[54]

Katherine Tucker, a visiting scientist at the Nutritional Epidemiology Program at Tufts University, noted that older adults lose the ability to absorb and use many nutrients. This natural aspect of aging means that the nutrients they do consume become less efficient, which in turn means that older adults actually need more nutrients than before. And although older adults may need fewer calories, they need more of certain types of nutrients and are advised to choose low-calorie, nutrient-rich foods. Research has linked fruit and vegetable consumption to the reduced risk of mortality among older adults. Older adults can mitigate chronic illnesses and overall poor life quality through healthful diet and exercise. This not only means freedom from debilitating symptoms and excess health costs, it can also extend the average life expectancy by 5 to 14 years.[55,56,57]

Maintaining good nutrition is significant for health and well-being. Although diet and lifestyle, combined with a healthy body weight, are important at any age, they are also crucial for healthy aging.[58]

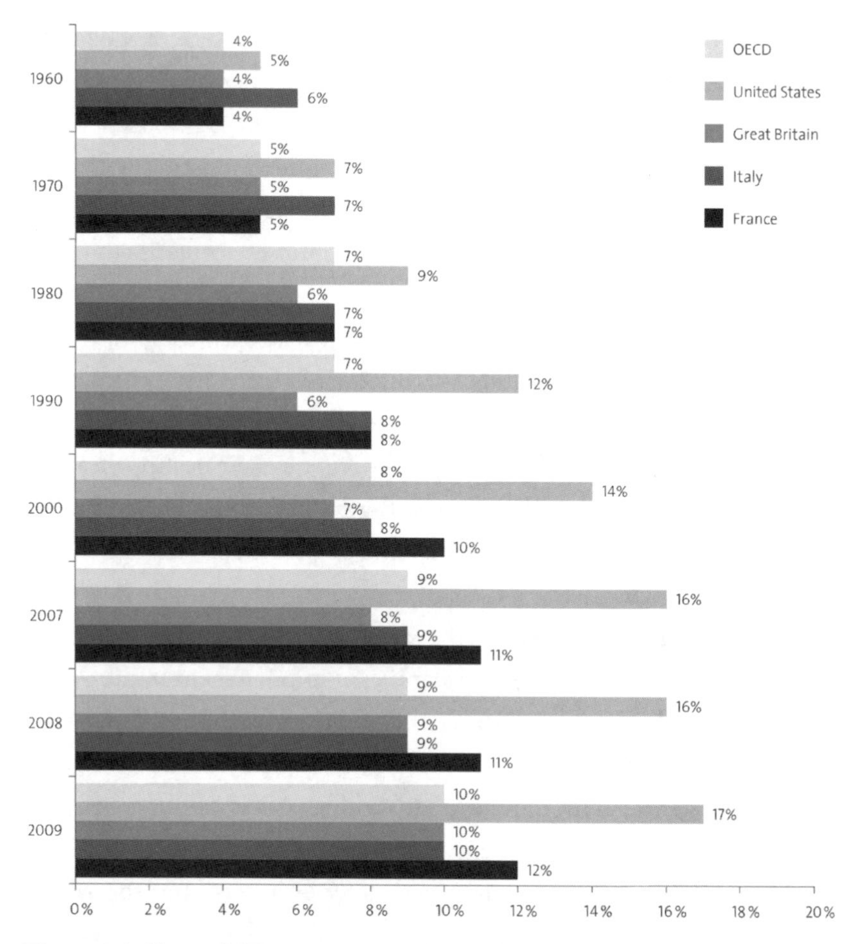

Figure 3.4. Share of GDP spent on total healthcare costs (1960–2009). Source: BCFN on OECD data, 2009, from p. 187 of *Eating Planet*, 2nd ed. (Milan: Edizioni Ambiente and the Barilla Center for Food & Nutrition, 2016).

Corporate Influence on Dietary Choices

The food industry is huge and complex—and often influenced by a focus on profit over health. While eaters rely on these businesses to make or distribute the foods they eat, these companies rely on influencing consumers to get the profits they need, regardless of cost or consequences.

According to a 2012 report from the Federal Trade Commission, U.S. food companies spent less than one half of 1 percent of their marketing dollars to promote fruits and vegetables, whereas the total they spent that year on food and beverage marketing to U.S. children was almost $2 billion. In 2016, the food and beverage industry spent more than $31 million lobbying Congress on issues related to healthcare and nutrition labeling on menus, so it clearly opposes regulation—either self-regulation or imposed regulation—on these issues.[59,60]

Corporations influence food choices most heavily through advertising, and the advertising of junk food to kids is the worst culprit of all in undermining public health. At an age when children are primed to learn the habits that will have lifelong effects on their health, these ads steer them toward unhealthful foods. Big Food is thus able to create a base of lifelong consumers at the expense of the health of children and the health of the future.[61]

A 2015 study in the *American Journal of Preventive Medicine* found that between 2007 and 2013, "no significant improvement in the overall nutritional quality of foods marketed to children has been achieved" by "industry self-regulation." The study concluded that "the lack of success achieved by self-regulation indicates that other policy actions are needed to effectively reduce children's exposure to obesogenic food advertising." Children and adolescents are now the target of intense and specialized food marketing and advertising efforts. Food marketers are interested in young people as consumers because of their spending power, their purchasing influence, and their potential as future adult consumers.[62,63]

Extensive meta-analysis reviews, such as those published in 2016 in the journals *Obesity Reviews* and *The American Journal of Clinical Nutrition*, have demonstrated how the sophisticated marketing techniques of the food and drink industries influence young people to adopt high-calorie, low-nutrient diets and contribute to their unhealthy weight gain, among other negative health outcomes. These studies have confirmed that food advertisers are successful not only at promoting one product over another but at increasing total consumption.[64,65]

Young children, especially under the age of 5, biologically prefer sweet and salty tastes, making them vulnerable to advertisements for their preferences. As children age, their ability to recognize branded logos increases, and, compared with other children, those who better recognize the logos of unhealthful products have a greater desire for these products. Adolescents between 12 and 18 years old are especially susceptible to online media and, by having more money to spend, are profitable targets for marketers of unhealthful foods and beverages.[66,67,68,69]

Anna Lappé, the author of several books about food and director of the Real Food Media Project, has cited Mixify as an example of Big Food advertising's tenacious focus on young people. This massive marketing campaign was introduced in 2015 by the American Beverage Association and several big beverage companies to deflect attention from sodas' contribution to ill health and to promote the idea that drinking sugary soda is okay if doing so is part of a balanced diet. The campaign included interactive social media accounts and a website peppered with teenage jargon, as well as free events for teens in cities across the country. Of the soda industry's obvious targeting of young consumers, Lappé said, "It's all about communicating that, dear (youthful) reader, we're hip, we're cool, we're one of you—and we don't want you to be worried about drinking our beverages."[70]

Marion Nestle, a professor of nutrition, food studies, and public health at New York University and author of the award-winning book *Food Politics*, has found that simply knowing certain foods pose health risks is not enough to encourage people to adopt a healthier lifestyle. "Environmental changes are much more likely to be effective," she said, "because education is aimed at changing personal behavior, which is too hard for most people to do. What you really want is to change the food environment to make it easier for people to make healthier food choices."[71]

Nestle notes that "eating less" and "eating better" cannot prevent obesity, because Big Food is determined to sell junk food, regardless of its effects on public health. As ways to counter the influence of marketers, she suggests stricter government regulations on advertisements and better education for consumers on how food is marketed. "People of every age are exposed to food advertisements all day long, so much so that food marketing has become part of the daily environment and is not consciously noticed," she said. "The objective of nutrition education clearly must be to teach critical thinking about food marketing in all its dimensions: advertisements, product placements in supermarkets, vending machines in schools, candy at the checkout counters of business supply and clothing stores, and cafes in bookstores. Noticing how food is marketed is the first step to learning how to resist it."[72]

A 2017 report by the Rudd Center acknowledged that the Children's Food and Beverage Advertising Initiative, an attempt by the Better Business Bureau to encourage industry to regulate itself, has made some progress. However, the report also confirmed that children's exposure to TV ads for some unhealthful categories, including candy, beverages, and fast-food restaurants, increased from 2007 to 2016.[73]

The fight against corporate influence on eating habits may seem at times like a losing battle, but there are small actions to take that can result in big differences. These actions begin in the home and in school environments, where children are growing and learning. It is crucial to teach children the habits of a healthy lifestyle, support programs that strive to regulate the power of Big Food, and educate eaters, young and old, on how Big Food markets its products without consideration for consumer health.

Box 3.2. A Healthy Balance between Eaters and the True Costs of Their Food

Agriculture is the base of food systems. In the view of Alexander Müller, study leader of The Economics of Ecosystems and Biodiversity in Agriculture and Food (TEEBAgriFood), "Agriculture is arguably the highest policy priority on today's global political agenda, in recognition of its widespread impacts on food security, employment, climate change, human health, and severe environmental degradation."[a]

Agriculture affects so many aspects of our society, and monitoring them with a standardized method can be a challenge for governments, businesses, and consumers. A deeper understanding of the effects provides a foundation for all actors to work together toward meeting the global demands for healthful foods and ethical and profitable sustainable farming practices.

A method being applied more widely in the food system today is an economic model called true cost accounting, which identifies and quantifies the total cost of food and agricultural production depending on the type of farming system. The goal is to improve the transparency of the true cost of producing food to increase fairness to farmers, make food more affordable for consumers, and decrease environmental and public health impacts.[b]

When making food choices, eaters may consider taste or nutrient content, but they are not often aware of the "true" cost of the food's availability at the market. The retail price of food rarely accounts for the cost of pollution, water scarcity, soil degradation, increasing GHG emissions, and poor labor standards, not to mention the health effects of nonnutritious foods.

To explain the concept of true cost accounting, Adrian de Groot Ruiz, executive director of the Dutch-based social enterprise True Price, compared conventional farming of Brazilian coffee in that country's Zona da Mata region with best-practices farming, which includes fair trade, organic, and agroforestry practices. He found that, on average, consumers were paying $2 per 2.2 pounds of conventionally farmed coffee. The true cost for that 2.2 pounds of coffee—factoring in the value of land use, water, energy, pollution, and fair

a. "Towards TEEBAgriFood," The Economics of Ecosystems and Biodiversity (2017), www.teebweb.org/publication/towards-teebagfood/

b. "True Cost Accounting," Lexicon of Food: Sustainable Food Trust (2017), https://www.lexiconoffood.com/new-food-systems/true-cost-accounting

and safe working conditions—was $5.17. In contrast, best-practices coffee cost consumers $2.78 per 2.2 pounds and had a true cost of $4.58.[c]

Ruiz suggested four possible interventions: Establish an equal pay program to address income discrimination, divert money from a community program to higher individual wages, increase the yields, and implement sustainable energy programs on the farms. He estimated that if farmers adopted these changes, not only would the true cost of 2.2 pounds of coffee drop to $3.79 in 5 years, but farmers who often operate at a loss would be truly profitable.[d]

Poorly functioning farming systems are not only detrimental to the environment, as discussed in Chapter 2, they can be harmful to the health of both eaters and farmworkers. In 2015, the United States reported more than 540 cases of cyclosporiasis, an intestinal illness linked to cilantro that's contaminated with human feces because workers in the Mexican fields where the cilantro is grown don't have access to bathrooms or the ability to wash their hands in the fields. And roughly 20 percent of all liver cancers are linked to maize contaminated by aflatoxin (a toxic mold caused by poor production and storage practices), particularly in sub-Saharan Africa.[e]

These alarming facts are the result of a broken food system, and the high, hidden costs should be catalysts for change and reasons to revolutionize current systems. Applying the method of true cost accounting to agriculture can provide a better understanding of the impacts—both positive and negative—that agricultural practices and food systems have not only on health but also on the environment, society, and the economy. In addition to measuring the negative costs associated with food production, true cost accounting can help quantify positive impacts of the food system on communities, society, and environment.

In 2016, Patrick Holden, president of the Sustainable Food Trust, organized the True Cost of American Food event in San Francisco, bringing together 100 of the most influential voices in food, including farmers, businesspeople, and

c. A. de Groot Ruiz, "True Cost Accounting in Practice," YouTube (2015), https://www.youtube.com/watch?v=EtPDUMRqUp4

d. Ibid.

e. "Cyclosporiasis Outbreak Investigations: United States, 2015," U.S. Centers for Disease Control and Prevention (CDC) (2015), https://www.cdc.gov/parasites/cyclosporiasis/outbreaks/2015/index.html; "Import Alert 24–23," United States Food and Drug Administration (FDA) (2017), https://www.accessdata.fda.gov/cms_ia/importalert_1148.html; F. Wu, "Perspective: Time to Face the Fungal Threat," *Nature* 516 (2014), https://www.nature.com/articles/516S7a

scientists, to discuss true cost accounting as a universal framework that could consistently and clearly measure the true cost of food.

Holden spoke of the importance of quantifying the externalities of food production: "How crazy, that if you want to make money in food or farming, the most profitable practice is to do things that are causing damage to the environment and the planet and our children and public health. While if you do the right thing, in terms of the planet and the people, you probably will make even less money or lose money. Unless you change that, we cannot expect farmers and food producers to make the change that is necessary."[f]

Another influential voice, Alexander Müller of TEEBAgriFood, said at the event, "It is important to understand that monetizing environmental impacts does not mean commodifying nature. The intent is to go beyond gross domestic product, the most powerful development figure in the world. But something must be fundamentally wrong with this indicator, as wherever there is damage to the environment or to people's health, GDP increases. . . . Sometimes, cheap food is very expensive, today or in the future, and true cost accounting aims to demonstrate this. For example, what are the costs of cheap food for the healthcare system?"[g]

When a corporation and the farmers it works with can share a focus on sustainability, these necessary changes can be made. A study by the University of Edinburgh looked at small-scale barley farmers in a water-scarce area of India to see whether best-practice advice from corporations could help them improve water efficiency, increase income, and reduce carbon dioxide emissions. The study found that these farm extension services could simultaneously create environmental benefits and raise the income of farmers while producing positive externalities.[h]

True cost accounting may be beneficial on many levels, although the numerous externalities involved with food production make it incredibly difficult to quantify costs. This challenge is compounded by the fact that the production

f. P. Holden, "The True Cost of Food," Sustainable Food Trust (2016), www.sustainablefoodtrust.org/articles/true-cost-food/

g. A. Müller, "Why We Need True Cost Accounting," in *The True Cost of American Food 2016 Conference Proceedings* (2016), http://sustainablefoodtrust.org/wp-content/uploads/2013/04/TCAF-report.pdf

h. C. Bowe and D. van der Horst, "Positive Externalities, Knowledge Exchange and Corporate Farm Extension Services: A Case Study on Creating Shared Value in a Water Scarce Area," *Ecosystem Services* 15 (2015): 1–10, https://doi.org/10.1016/j.ecoser.2015.05.009

of different food categories (e.g., vegetables versus meat, processed food versus fresh food) varies significantly.

Trucost, which focuses on informing business and investment decisions by incorporating environmental and climate data, examined the stages of production, from farm to supermarket shelf, of three common food products: breakfast cereal, fruit juice, and cheese. The embedded carbon, water, waste, and pollution were calculated for generic products in each category. Trucost then calculated the "natural capital" cost (i.e., the impacts to water and air pollution) of each of these. Its analysis indicated that, on average, the true cost of a block of cheese should be 18 percent higher than the average retail price, breakfast cereal should be 16 percent more expensive, and fruit juice 6 percent more.[i]

Aine Morris, a founding member of Slow Food's Youth Food Movement, remarked on how unaware most consumers are about the true cost of food, noting that "the problem is that the additional costs are often less visible than the direct, tangible impact we feel on our pockets when buying the weekly shop. We urgently need to ensure that the real, long-term cost of the way our food is produced is visible to the consumer, and true cost accounting can help us do this." Morris noted that, rather than raising retail prices to cover the true cost of a product, true cost accounting facilitates the administration of incentives, taxes, subsidy redistribution, and policy initiatives to promote the expansion of sustainable farming practices and to ensure higher production costs for those damaging the planet or human health. She cited taxing nitrogen fertilizer as a way to encourage farmers to use it conservatively, and she suggested contributing the tax funds to an educational program promoting planting legumes as an ecofriendly way to add nitrogen to the soil.

A report on true cost accounting from the UN Food and Agriculture Organization expressed the importance of a standard method to be used by all: "Quantifying the full costs of food wastage improves our understanding of the global food system and enables action to address supply chain weaknesses and disruptions that are likely to threaten the viability of future food systems, food security, and sustainable development."[j]

---

i. J. Hollender, "Full Cost Accounting," Jeffrey Hollender Partners (2017), www.jeffreyhollender.com/full-cost-accounting/

j. "Food Wastage Footprint: Full-Cost Accounting," Food and Agriculture Organization of the United Nations (2014), www.fao.org/3/a-i3991e.pdf

The Food Business Can Do Better

The way businesses in the food sector operate has an impact not only on the environment but also on global health. When business leaders have a better understanding of a company's dependence on natural resources and the impact it has on those natural resources and society, they can change practices to minimize negative externalities and promote positive ones. They can also create models that respond to consumer demand for healthier food and more sustainable food systems while creating a transparent relationship with processors and consumers.

In 2014, Tom McDougall founded 4P Foods, a community-supported food delivery service in Washington, D.C., that sources from local farms and makes fresh produce more accessible. 4P Foods' business model proved, through its own success, that businesses can make a profit and also be environmentally and socially responsible. McDougall's model calls for companies to take responsibility for their impacts on the people who grow the food, the people who eat the food in the community, and on the planet. He focuses on the four P's of sustainable agriculture: purpose, people, planet, and profit. "The businesses of today are creating the world of tomorrow," said McDougall. "They have the ability and, more importantly, the responsibility to use their economic engines to drive real, lasting, positive social change."[74]

Like their counterparts in any other economic sector, companies in the food industry respond to the needs and preferences of consumers to ensure that there continues to be a market for their goods and services. However, determining how to be both profitable and environmentally sustainable can be a challenge. According to the Sustainable Food Trust's Patrick Holden, "The absence of an enabling economic environment for sustainable food production represents the single biggest barrier to mainstreaming the transition toward sustainable food systems."[75]

Therefore, eaters, along with policymakers, the funding and donor communities, and other decision makers can dissolve this barrier and encourage that true cost accounting be used throughout the food and economic system. This would push companies to make changes to their product lines, policies, and actions and to integrate sustainable systems into the mainstream.

Businesses in the food sector, especially restaurants, can be very responsive to consumer voices. For example, Chipotle is going genetically modified organism (GMO) free, citing concerns about the impact of GMOs on the environment, farmers, animals, and their customers. McDonald's and Chick-fil-A have pledged to become antibiotic-free, and because of concerns about antibiotic resistance, Tyson will eliminate the use of human antibiotics in chicken feed by the end of 2017. Major grocery chains have committed to sourcing organic food; Costco became the largest organic grocer in the United States in 2015, and Kroger topped $11 billion in sales of natural and organic foods in the same year. By their sheer force and size, these companies can change the way the food system works.[76,77,78]

Environmental expert D.A. Bainbridge, a professor of environmental studies at Alliant International University, advocates for nongovernment organizations (NGOs) and farmers themselves, as well as companies, to step up and demand the implementation of true cost accounting. "Research on environmental and social costs is desperately needed, and companies and NGOs will have to play a major role in undertaking and funding needed research because governments, constrained by budget problems from unsustainable decisions in the past, are not likely to do so," he said. "Special interests have crippled and controlled environmental and social cost research in recent years, which has led to efforts by parts of the U.S. government to deny the effects of global warming, acid rain, social inequity, health risks, and many other problems."[79]

Using the true cost accounting framework, all actors along the food supply chain can understand the impacts of their practices and then use that information to adopt better ones. This can provide the structure businesses need to be both profitable and environmentally sustainable. At the same time, the changes businesses can make with true cost accounting can help build an environment in which consumers know more about the food they eat and can more easily make healthier, more sustainable choices. And healthy eaters mean a healthier society as a whole.

When those in the agri-food sector apply the economic model of true cost accounting, they can help the planet and continue to profit. More important, however, is that businesses apply changes to their practices that can improve the health and social strength of consumers and provide better assistance to food and food access.

The Center for Science in the Public Interest (CSPI), which calls itself "America's Food Watchdog," is a consumer advocacy organization focused on conducting innovative research and advocacy programs in health and nutrition and providing consumers with current, useful information about their health and well-being. CSPI regularly confronts corporate food actors on behalf of consumers. For example, in 2016 CSPI filed lawsuits against PepsiCo and the U.S. Food and Drug Administration (FDA); won a concession in an ongoing lawsuit against the FDA; settled a lawsuit against Vitaminwater and its parent, Coca-Cola; won a campaign to get Mars to remove dyes from M&M's candies; and filed regulatory petitions with the U.S. Department of Agriculture.[80,81]

The Interagency Working Group on Food Marketed to Children has a similar mission. This initiative of the U.S. Department of Agriculture, the FDA, the U.S. Centers for Disease Control and Prevention, and the U.S. Federal Trade Commission was created in the early 2010s to propose nutrition principles to guide industry self-regulatory efforts.[82]

Few realize the role agribusiness plays in eater health. As consumers shop and go about their daily lives, Big Food is a constant influence through its media presence and advertisements. Profit-oriented corporations may not always respond to self-regulation guidelines; however, their customers have the power to demand that companies change. Along with the efforts of NGOs and health advocacy organizations to regulate Big Food on the government level, change for a better, healthier food system is possible.

Conclusion and Action Plan

As much as health is a choice, it can at times seem like a challenge to make healthful decisions. The creation of lifelong healthy eating habits comes at a tenuous time, during childhood, when we are most open to learning long-term behaviors. Should the nutritious food central to good health be unavailable or, by advertorial influence, avoided, a lifetime of poor health becomes a likelier future.

The health problems that accompany poor dietary choices and malnutrition are not limited to either hunger or obesity; both issues create a paradoxical problem experienced—at times simultaneously and even in the same locations—by communities worldwide. Across the globe, populations both young and old, rich and poor, will suffer the effects of this paradox. However, the scales have tipped in the past few decades toward obesity, and the number of obese people worldwide now outnumbers the population of the hungry and malnourished who were once the majority. The surge in obesity can be attributed, in great part, to the billions of dollars spent by large agribusinesses on heavily targeted advertising and on lobbyists to turn the decisions of Congress in their favor.

The health of the future is specifically undermined by food and beverage advertisements focused on young eaters. According to obesity expert Aviva Must, Big Food has a major role to play in determining the dietary habits of children and adolescents. Developing pro-health actions for the industry to make can come with roadblocks. "Unfortunately, there are economic disincentives to many of the best ideas," Must said. "One would like to see the industry make a business commitment to health—elevating the manufacture of healthful products to be a key criterion for their activities."[83]

Good health today, and good health for the future, relies largely on the health of our food systems. All costs—from the environmental cost of food production to its retail cost and the cost to eaters' health—should be considered when choosing foods, yet so often they are not. The application of a structured true cost accounting approach can help evaluate the social, environmental, economic, and health-related costs of food and can benefit food systems by providing governments and businesses with tools they can use to improve decision making.

The odds may seem stacked when it comes to ensuring good health through nutritious foods for all. Despite the challenges posed by Big Food and the lack of a standardized true cost accounting method, hardworking health advocates are making positive changes to our global food system.

The good news is that family farmers, food heroes, and organizations around the world are working together to improve access to nutritious food and to develop more sustainable local food systems that can be resilient in the wake of ecological disturbances and foster the health of growers and eaters.

One such organization, the World Food Programme (WFP), has set its sights on achieving Goal 2 of the UN Sustainable Development Goals: eliminating hunger entirely by 2030. Importantly, the WFP is including micronutrient deficiencies and other forms of "hidden hunger" in its approach to treating and preventing malnutrition. By building capacity within critical countries, facilitating both cash and in-kind food assistance, promoting south–south cooperation, and remaining prepared for emergency relief operations, the WFP has made itself the world's largest humanitarian organization focused on addressing hunger and food security.[84,85]

Africa Rising is a development partnership that creates opportunities for smallholder farm households to improve food, nutrition, and income security, and to conserve natural resources, through sustainably intensified farming systems. La Red de Guardianes de Semillas (The Network of Seed Guardians) is preserving rare plant varieties and culturally important seeds in Tumbaco, Ecuador, with a goal of training growers about how permaculture works and preserving biodiversity throughout Ecuador.[86,87]

Among the most prominent of these efforts is the UN Environment Programme's initiative known as The Economics of Ecosystems and Biodiversity in Agriculture and Food (TEEBAgriFood). This team of farmers, economists, business leaders, agriculturalists, and biodiversity and ecosystem experts is reviewing the economic interdependencies between human systems, such as irrigation, labor, agriculture, and food production, and natural systems, such as biodiversity and ecosystems. TEEBAgriFood's work has included a series of sector-specific, geographically widespread "feeder studies" to assess the hidden environmental and social costs of different agricultural commodities, such as rice, livestock, palm oil, inland fisheries, maize, and agroforestry, and the organization will release a new report on the true costs and benefits of food production and consumption in 2018.[88]

Achieving and maintaining good health, for both eaters and the environment, really is a team effort, especially when we consider the influence of Big Food on eater habits. While pro-health organizations advocate for eaters and push back on government policies that favor Big Food, consumers have the power to make more positive personal changes. The biggest impact they can have starts in their own homes and their own communities, by encouraging the healthy habits of children and family members and by urging schools and local governments to focus on better food for health.

Eaters can take control of their health and their future by

- Encouraging good behaviors and lifestyles early in a child's life.
- Monitoring a child's exposure to food and beverage advertising.
- Maintaining a healthful diet and an active lifestyle throughout the child's life.
- Recognizing the "true costs" of foods and trying to make the best food choices possible for their own health, the health of food producers, and the health of the planet.

VOICES FROM THE NEW FOOD MOVEMENT: Alexander Müller

What has made you stay involved in food and agriculture for so many years?

My answer is twofold. The political side is that the way we are producing food is one of the most important and pressing issues for sustainability and human wellbeing. It has major positive or (unfortunately very often) negative impacts on natural resources, it shapes the landscape worldwide, it generates income for billions of people, and it is linked with knowledge, education, social equity, and global justice. If we do not transform today's agriculture into real sustainable food systems, we will not achieve the UN Sustainable Development Goals (SDGs), and we will not eradicate poverty and inequality. It is a truly cross-cutting issue and is important for everyone.

At a personal level, having dinner with friends and even people you do not know and sharing a bottle of wine is one of the things I enjoy a lot. It shows the importance of food for human relations and our culture.

What do you see as the biggest opportunity to fix the food system?

The fact that agriculture is now embedded in the framework of the SDGs. Achieving sustainability goes beyond vested interests, which are very strong in the agricultural sector. The commodification of agriculture is not a way towards sustainability!

What innovations in agriculture and the food system are you most excited about?

Access to information and the new possibilities of decentralized information systems allow communities on the ground to develop their business and get access to markets. In addition, consumers can obtain information about their products. We need a system that is independent and works in a transparent way.

Can you share a story about a food hero that inspired you?
Tewolde Berhan Gebre Egziabher from Ethiopia won the Right Livelihood Award in 2000 "for his exemplary work to safeguard biodiversity and the traditional rights of farmers and communities to their genetic resources."

For the past 30 years, Tewolde has been a key driver to overcome hunger and starvation in Tigray, an area of Ethiopia where—according to reliable sources—in the 1980s and 1990s, millions of people were starving or hungry. He initiated the Project Tigray to demonstrate that food security could be better achieved by building on farmers' traditional knowledge, adapting to available local resources, and creating jobs. He introduced the use of compost, a new technology to most of the farmers in the region.

As a result, the government adopted a policy guideline based on his success: "Ensure that essential ecological processes and life-support systems are sustained, biological diversity is preserved, and renewable natural resources are used in such a way that their regenerative and productive capabilities are maintained, and, where possible, enhanced." What a modern and forward-looking guideline!

What drives you every day to fight for the bettering of our food system?
We also have to eat every day!

What's the first, most pressing issue you'd like to see solved within the food system?
Access to food for everyone! The world produces enough for everybody, but we are wasting around 30 percent of all food produced. What a scandal.

What's one issue within the food system you'd like to see completely solved for the next generation?
Hunger and malnutrition. But based on sustainable systems and not land degradation and destruction of biodiversity.

What agricultural issue would you like for the next president of the United States to immediately address?
Linking nutritional guidelines to sustainability.

Alexander Müller is a study leader of the UN Environment Programme–hosted project TEEBAgriFood. Previously, he served as state secretary in the Ministry for Consumer Protection, Food, and Agriculture in the Federal Republic of Germany and in the position of assistant director general of the Food and Agriculture Organization of the United Nations.

Citation: "Ten Questions with Alexander Müller, Study Leader of TEEBAgFood," Food Tank (2016), https://foodtank.com/news/2016/04/ten-questions-with-alexander-mueller-study-leader-of-teebagfood/. *Interview edited by Michael Peñuelas in August 2017.*

VOICES FROM THE NEW FOOD MOVEMENT: Bruce Friedrich

What inspired you to get involved in food and agriculture?
In 1987, I read *Diet for a Small Planet* by Frances Moore Lappé. Before then, it had somehow not occurred to me that farm animals have to eat a lot more calories in feed than they turn into meat. It's a vastly inefficient system that wastes at least 90 percent of the caloric inputs (and that's just for chicken—it's even worse for other animals). Some of the animals' feed they burn off simply existing (just like everyone reading this eats more than 1,000 calories a day just to exist), and some of that feed produces inedible parts of the animals—bones, blood, fur, etc.

Most of us recoil at wasting food, but the waste involved in feeding grains and soy to animals that they simply burn off is just as real as if we threw all the food straight into the trash. I open with this concept and expound upon it in a presentation I gave at a recent Bread for the World conference.

A few years back, the United Nations (UN) Special Rapporteur on the Right to Food argued that biofuels are a human rights crime because they divert 100 million metric tons of corn and wheat to fuel, driving up the price of those crops and leading to starvation. That same UN report he was citing indicated that 756 million metric tons of corn and wheat are diverted to feed farm animals, with equally dramatic economic effects. And that doesn't even count the 85 percent of the global soy crop that is fed to farm animals, leading to scarcity, rain forest deforestation, and the displacement of farmers in the developing world.

What do you see as the biggest opportunity to fix the food system?
At The Good Food Institute, we see plant-based and "clean" meat (i.e., meat produced in a culture, without animal slaughter) as most likely to be maximally transformative. Animal agriculture is a major contributor to climate change and global poverty, produces products that are generally unhealthy, and supports a treatment of animals that most people find deeply objectionable. These are points that are generally accepted by anyone who considers them. So it's clear that conventional animal agriculture is ripe for innovation, and we exist to encourage that.

What innovations are you most excited about?
We're very excited about a variety of innovations in both plant-based and clean meat, dairy, and eggs. In particular, we're excited about what Beyond Meat, Impossible Foods, Hampton Creek, and Memphis Meats are doing.

For example, Beyond Meat, which counts Bill Gates among its investors and boosters, just debuted the first raw, plant-based meat available right in the meat counter, and it sold out in hours. Impossible Foods, which has received (and turned down) buyout offers from Google to the tune of hundreds of millions of dollars, is bringing the most realistic plant-based burger to market imminently. Hampton Creek is doing amazing work with plant proteins and plans to create an open source database to accelerate even more segment growth—similar to what Elon Musk has done with electric car technology. And Memphis Meats has raised millions of dollars for clean meat development; *Fortune Magazine* called them the hottest tech in Silicon Valley.

There is a lot of other incredibly exciting work happening in both plant-based and clean food technology, and it would take pages to do the question full justice. The future for these technologies and the companies taking on this pioneering work is insanely bright.

Can you share a story about a food hero that inspired you?

I can't pick one food hero—there are just too many. Really, I'm inspired by anyone who thinks about new ways to solve big problems. I'm inspired by anyone who sets goals that are focused on making the world kinder and our economy more sustainable. For example, I'm inspired by Bangladeshi entrepreneur and banker Muhammad Yunus and the bank he founded, Grameen Bank; R. J. Pachauri and the Intergovernmental Panel on Climate Change (IPCC); Elon Musk and Tesla Motors; and everyone who is working to bring cell phone technology to the far reaches of the Earth to bring weather reports, crop prices, and banking to subsistence farmers. I'm especially inspired by the venture capitalists who see a problem and choose to fund the innovation that will solve the problem. In the food space, specifically, I'm inspired by Bill Gates' support for Beyond Meat and other plant-based food tech companies, and by Li Ka-Shing and Sergey Brin because of their support for clean alternatives to animal-based meat, dairy, and eggs. I am also inspired, of course, by the food innovators themselves, and probably none more so than Memphis Meats' CEO Uma Valeti, who left a successful cardiology practice at the University of Minnesota to put all of his energy into creating real meat that doesn't have all of the global health impacts and threats that conventional animal meat has. Dr. Valeti is a visionary, and he's probably my top food hero. His commitment is deeply inspiring to me.

What drives you every day to fight to better our food system?

Right now, plant-based milk comprises more than eight percent of the milk market, but plant-based meat comprises only about 0.25 percent of the meat market. I have no doubt that we can close and then surpass that market share, and it's exciting to me to think about the fact that simply getting plant-based and clean meat to 15 percent would spare about 1.4 billion land animals per year and more than 2 billion sea animals. It would also have massive positive impacts on global poverty and other sustainability issues, climate change and other environmental issues, and global health.

Just about everyone wants to make choices that are consistent with their values, so just about everyone supports sustainable farming practices, wants to lessen their adverse impact on the climate, and wants to see animals treated well. The solution to a lot of very big problems is pretty simple—we just need to create and promote the companies that are making plant-based and clean alternatives to animal products a reality. That's why we formed The Good Food Institute (GFI)—to accelerate the plant and cultured spaces as effectively and efficiently as possible.

What's the biggest problem with the food system that our grandparents didn't have to deal with?

Our grandparents didn't live in the same global food system that we do, and meat consumption was a fraction of what it is now. They were dealing with far fewer animals and had not heard of climate change. Plus, the system wasn't as globalized: there wasn't the same link between eating meat in one country and deforestation and poverty elsewhere.

What's the first, most pressing issue you'd like to see solved within the food system for the next generation?

A key focus of GFI is to put plant-based and clean alternatives to animal products onto the radar and into the budgets of governments, big food companies including meat companies, and venture capital investors so that we can address a variety of issues and opportunities at once. There is tremendous financial opportunity in plant-based and clean alternatives to animal products, and these products can also help governments deal with climate change, resource scarcity, and the prospect of a world without antibiotics; basically, we're talking about solving some of the world's biggest problems by using a clear solution that is currently vastly under-resourced. On the one hand, these are massive issues; on the other hand, the solution is pretty simple—markets and food technology.

I'm not sure we can solve it for the next generation, but I don't think we're too far away from a world in which 100 percent of meat is either plant-based or clean (i.e., grown in a culture without animal slaughter). We have the technology—we just need the will. Considering the myriad of problems in conventional animal agriculture, I think it's likely that, in a few generations, animal slaughter for food will be extremely rare in the developed world.

What is one small change every person can make in their daily lives to make a big difference?

Obviously, people can eat less meat—that's pretty easy. But much more important is for people to educate their friends, family, key policymakers, corporate executives, and so on. If someone is on the Food Tank website, they're off to a good start. It's important to find the causes with regard to food that excite you and to get active. Changing our personal decisions is great, but real power resides with changing as many others as possible, and then the entire system. Supporting The Good Food Institute is a great place to start, in my opinion.

What agricultural issue would you like for the next president of the United States to immediately address?

Subsidies. It's ridiculous that foods that contribute to such an array of problems are subsidized by the U.S. government (and most other governments, of course). Conservatives tend to oppose subsidies because they are anticompetitive and interfere with the fair workings of the marketplace, and liberals tend to oppose subsidies to industries that are harming our environment and human health. But because of the power of animal agriculture, the subsidies persist. It's time for a president to spend some political capital to get rid of subsidies for products that are bad for the environment, our health, and animals.

Bruce Friedrich is executive director of The Good Food Institute, which works with scientists, investors, and entrepreneurs to create and promote clean meat and plant-based alternatives to animal products. He is also a co-founder of New Crop Capital, a $25-million venture capital fund that invests in companies that are producing alternatives to industrial meat, dairy, and eggs.

Citation: "Ten Questions with Bruce Friedrich, Executive Director at The Good Food Institute," Food Tank (2016), https://foodtank.com/news/2016/10/ten-questions-bruce-friedrich-executive-director-the-good-food-institute-2/. *Interview conducted by Kate Reed in October 2016 and edited by Michael Peñuelas in August 2017.*

VOICES FROM THE NEW FOOD MOVEMENT: Tristram Stuart

How did you develop the idea for Feeding the 5000?

Feeding the 5000 was born from my work on my book, *Waste*, and the organization I founded, Feedback. What animated it was a lot of joint thinking about how to demonstrate in a really visual, visceral way just how much food is wasted the world over and to show businesses that people really care about this problem. I also wanted to create something that used the power of food to bring people together to celebrate the fact that although food waste is a massive problem, the solutions can be positive—and delicious! There have been over 40 events worldwide, and on May 4 Feeding the 5000 is coming to Los Angeles, California for the first time. We will be partnering with L.A. Kitchen and several city departments who have been working hard on tackling how much waste the city sends to landfill through Zero Waste LA.

Can you describe some of the past events? What do people learn when they eat food that has been recovered?

It all started in London in 2009, and from there Feedback has worked with local organizations to produce Feeding the 5000 events from Paris to Dublin, then even further afield, to Sydney, New York City, and Washington, D.C. What's palpable at each event is the power of a good meal to bring people together and get them talking to one another about the massive impact of all this waste on the environment. Sharing this meal really brings home that what is being wasted isn't rubbish—it's good, nutritious food that should be filling bellies, not bins. We want people to go beyond feeling shocked or guilty about food waste, to ask questions about why waste is occurring on this scale, and sign Feedback's pledge to waste less themselves and ask businesses to do the same.

Why did you choose to bring Feeding the 5000 to Los Angeles a few years ago?

Being in Los Angeles will be a wonderful and bizarre mix of showcasing the incredible work of our partners on hunger, food, and the environment against the backdrop of L.A.'s reputation as a symbol of excess amid a highly climate-conscious state that's facing huge environmental challenges. When you consider that California supplies 90 percent of some vegetables eaten in the rest of the country, like broccoli, but then up to 40 percent of all that food is wasted, it makes you realize that states like California are bearing a disproportionate burden of the climate and environmental impact of food waste in the U.S. We worked out that 9,000 million gallons of water per day in California are used to produce food that is never eaten—that's equivalent to leaving a tap running all day every day for 9,000 years. That's a serious amount of water down the drain because of food waste.

Can you talk a bit about Toast Ale and the process for turning recovered bread into beer? What has been the response from consumers?

I started Toast Ale in 2016. It is a delicious craft beer made with surplus fresh bread, designed to help both highlight and tackle the fact that bread is one of the most frequently wasted items in the U.K. Until Toast Ale, there has not been a scalable solution to surplus bread: it is overproduced in such massive quantities that soup kitchens and food recovery groups routinely turn away fresh, unsold bakery loaves. Beyond preserving valuable, resource-intensive grains for human consumption, 100 percent of the profits from Toast Ale will be poured into Feedback and other charities tackling the root causes of food waste. We have just launched a crowdfunding campaign on Indiegogo to bring production to the USA as well.

What plans does Feedback have for the future?
We're currently calling on food businesses, particularly supermarkets, to be transparent about how much they waste and where waste occurs. Without that transparency, you're really working in the dark when you are looking for ways to prevent waste, like Toast Ale. All our programs focus on one goal—building a global food system that is sustainable and fair, and doing it through challenging power and inspiring people to achieve change.

Tristram Stuart is the founder of the organization Feedback and the author of *Waste: Uncovering the Global Food Scandal.* Feeding the 5000 was one of Feedback's first initiatives and took place in December 2009 in London's Trafalgar Square. The event fed 5,000 people using only ingredients that would otherwise have been thrown out for being cosmetically imperfect. Since 2009, Feedback has continued to grow as an organization, and Feeding the 5000 events have been held worldwide.

Citation: "Tristram Stuart Is Fighting Food Waste, 5000 Meals at a Time," Food Tank (2017), https://foodtank.com/news/2017/05/interview-feedback-founder-tristram-stuart/. *Interview conducted by Zoya Teirstein in 2017 and edited by Michael Peñuelas in August 2017.*

CHAPTER 4

Food for Culture

THE CULTURE OF HOW, WHY, AND WHAT we eat is complex and varied, changing from country to country, region to region, village to village, and even household to household.

Cooking with fire, according to anthropologist Claude Lévi-Strauss, is "the invention that made humans, human." Indeed, before we learned how to cook, we ate food, meat in particular, that was raw, rotten, and even putrefying. The use of fire was a turning point in history. Cooking food—transforming it into something more easily edible and certainly more palatable—marks the transition between *nature* and *culture* and also between *nature* and *society*. Cooking creates links that otherwise would not have come about; it creates social and cultural norms, and it can create generosity. According to good-food advocate and author Michael Pollan, "Cooking is all about connection, I've learned, between us and other species, other times, other cultures human and microbial both, but, most important, other people. Cooking is one of the more beautiful forms that human generosity takes. . . ."[1]

In addition, says psychologist Paul Rozin, national cuisines embody the dietary wisdom of populations and their respective cultures. It is not far-fetched to argue that the history of humanity's relationship with food has been an extraordinary social and cultural saga of a quest for meaning, in which even the most problematic aspect—searching for food to nourish oneself—was transformed into the opportunity to evolve as the human race and create the civilization we see today.[2]

Food represents a multitude of meanings, and it can serve both as a form of social interaction and as a communication tool. Food provides a foundation for conveying religious norms, but it can also support the expression of human identity and sexuality and be the ground on which conflicts and control play out.

At the same time, food has led to shifting power dynamics between women and men, religions, competing political viewpoints, and different cultures. Food is more than just vitamins and nutrients; it provides the ties that bind communities, economies, and the environment. This chapter focuses on the importance of preserving and improving both agricultural and cultural diversity, the need to recognize the world's great culinary traditions, and why we need a new vision of food and nutrition for the future.

Going Forward by Going Back

In 1986, Carlo Petrini did something no one had ever done before. He protested the opening of a McDonald's restaurant in Rome, Italy. At that time, Petrini was a journalist and, although he didn't call himself one, a foodie. To demonstrate his objection to the fast-food chain, he handed out plates of penne pasta, reminding Italians and tourists alike that Italy is a land and a culture based on the Mediterranean diet, which focuses on diversity—colorful vegetables, legumes, fish, grains, olives, and nuts, among other foods—and serves as the base of Italy's world-renowned food.[3]

What Petrini feared about fast-food culture was its effect on the diversity of foods we eat. "I was alarmed by the culturally homogenizing nature of fast food," he said in *Time* magazine.[4]

Lack of diversity is just one problem that results from the loss of traditional diets and cuisines. Every day, plant species across the globe are disappearing. The UN Food and Agriculture Organization (FAO) reports that approximately 75 percent of Earth's plant genetic resources are now extinct, and another third of plant biodiversity is expected to disappear by 2050. Up to 100,000 plant varieties are currently endangered worldwide.[5]

Unfortunately, most investments in agriculture from research institutions and the funding and donor communities are for commodity crops such as wheat, rice, and maize rather than for more nutritious foods or indigenous crops, and this focus has had devastating consequences. Lack of diet diversity contributes to undernutrition and overnutrition.[6,7] Global obesity rates have doubled over the past 30 years, and the incidence of diet-related illnesses, including diabetes, hypertension, and heart disease, in industrialized and developing countries alike, has exploded.[8]

All this despite the fact that many indigenous foods are environmentally sustainable, improve food security, help prevent malnutrition, and increase farmers' incomes. Fortunately, though, across the world there are dozens of initiatives helping to preserve ancient flavors and promote better nutrition.

The same year Petrini made his McDonald's protest, he started the organization Slow Food International. Today, some 32 years later, Slow Food has more than 150,000 members across 50 countries.[9]

The organization's Ark of Taste initiative is cataloguing indigenous species of fruits and vegetables all over the world. And every 2 years, Slow Food brings members together to celebrate Terra Madre, or Mother Earth, at Salone del Gusto in Turin, Italy. There, 10,000 Slow Food farmers share both their products with attendees and their obstacles, successes, and best practices with one another through meetings and workshops.[10]

CHOOSE FOODS CONSCIOUSLY

Humans have remarkable capacities for recognizing and memorizing, and these skills help people to avoid poisons and to find the most nutritious foods. Aside from their senses and memories, individuals base their food choices on culture and traditions that preserve the flavor and experience of countless "tasters" who went before them

THE OMNIVORE'S DILEMMA

Culture codifies the rules of a wise diet with a complex series of taboos, rituals, recipes, regulations, and traditions. All of this allows human beings to avoid being faced on a daily basis with "the omnivore's dilemma"

GREATER FAIRNESS IN THE WORLD

Fairer food means that we have a responsibility for our weaker neighbors, that we value food as a means of peaceful coexistence among peoples, and that we find ways to establish socio-economic equilibriums through the phases of production

Figure 4.1. Food for culture.

REDISCOVERING THE PLEASURE OF FOOD

The great challenge of our time is to redevelop a deeper, richer, more meaningful relationship with food, where the relationship with the things we eat is restored to the dimension of esthetics, taste, and conviviality

FIGHTING OBESITY AND FOOD-BASED PATHOLOGIES

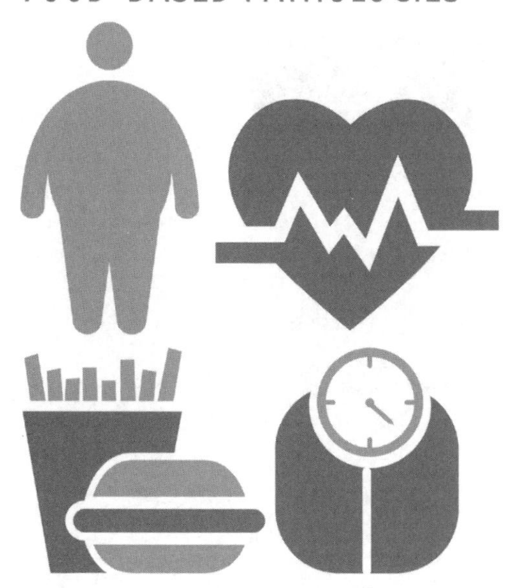

Eating has become a banal experience, leading to the epidemic of obesity and diseases linked to obesity

THE IMPORTANCE OF CULINARY TRADITIONS

Currently, we are witnessing the progressive abandonment of the gastronomical traditions of the past, as well as the loss of knowledge about cooking and the makeup of food

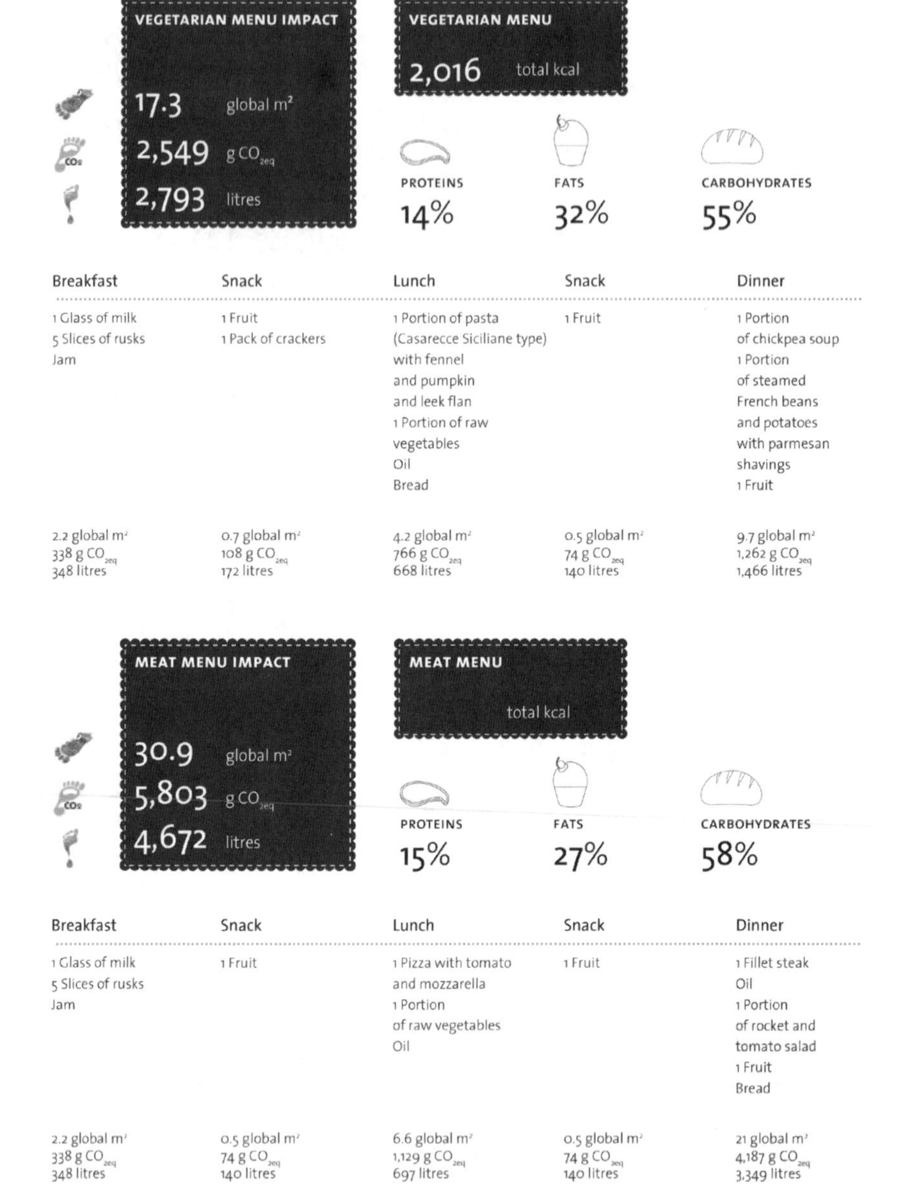

Figure 4.2. Footprint and food choices. Source: BCFN, 2015, from p. 105 of *Eating Planet*, 2nd ed. (Milan: Edizioni Ambiente and the Barilla Center for Food & Nutrition, 2016).

According to Slow Food vice president Edie Mukiibi, the young farmers he works with in Uganda and Kenya face numerous challenges. But they're also making great progress toward solving these problems, in part by growing the foods their grandparents and great-grandparents grew. "You find little children of 3 to 15 years having a lot of knowledge about the traditional crops, the local crops, the planting seasons, and such kinds of things," Mukiibi told Voice of America. "This is what we are achieving with the gardens."[11]

The gardens he's referring to are part of Slow Food's 10,000 Gardens in Africa Project, which aims to encourage communities all across the continent to grow indigenous crop varieties—crops that are not only resistant to pests and disease, drought or flooding, or poor soils but are tasty.[12]

One good example is amaranth, a leafy vegetable that grows quickly in the humid lowlands of Africa and that is traditionally consumed in countries including Togo, Liberia, Guinea, Benin, and Sierra Leone. Amaranth thrives in hot weather and is an excellent source of protein, vitamins, and essential minerals, including calcium, iron, magnesium, potassium, and zinc. Or the cowpea, which originated in Central Africa and is one of the region's oldest crops. It is also drought-resistant and can thrive in poor soil conditions, and even its leaves, not just its protein-rich peas, are nutritious.[13] Or the spider plant, a green leafy vegetable, also known as the African cabbage, which flourishes throughout Africa. It is high in protein, antioxidants, vitamins, and other micronutrients.[14]

Selected Endangered Foods Worldwide and Efforts to Save Them

Mandaçaia bee honey, Brazil

Special Characteristics: The honey is produced largely by women as a source of income. Each hive can produce 1 to 1.5 liters of honey a year; the honey is used as a food or as medicine to treat breathing problems, coughs, and flu.[15]

Why Endangered: Threats include the use of insecticides on fruit crops, droughts, and the unsustainable harvesting of honey. Also, umburana trees, which the bees use for nesting, are being cut down for timber.[16]

Efforts to Save: Slow Food is teaming up with local beekeepers to safeguard the bees from extinction by promoting the exchange of knowledge of sustainable practices. Selling the honey to local restaurants provides an incentive to conserve the bees.[17]

Maungo, Angola

Special Characteristics: The larvae of the maungo caterpillar are an important protein source for rural communities. They are collected during the rainy season and sun-dried. The dried larvae are then sold in local markets.[18]

Why Endangered: The number of caterpillar larvae has been limited by decreased rains resulting from climate change. Local communities, looking to supplement the income loss caused by low larvae yield, cut down mopane trees, where the caterpillars live, to make charcoal.[19]

Efforts to Save: The maungo and diverse other species depend on the mopane forest of southeastern Africa. The World Wildlife Fund is aiming to conserve more than 45% of the region's mopane forests and plans to expand these state and private conservation measures to southwestern Africa.[20]

North American ginseng, United States and Canada

Special Characteristics: Ginseng, which was used by Native Americans to treat headaches, indigestion, and fever, is now used for diabetes care, tumor reduction, immune system improvement, and treatment of attention deficit–hyperactivity disorder.[21]

Why Endangered: The plant, which takes 6 years to fully mature in the wild, is largely overharvested. There is great demand in

Asia, where native ginseng has been overharvested, which adds pressure.[22]

Efforts to Save: In Quebec, 38 conservation sites and programs were established in 2004–2005 to offer scientific support to nongovernment organizations' monitoring and conservation efforts.[23]

Bunya nut, Australia

Special Characteristics: From December to March, bunya pine trees drop soccer ball–sized cones containing edible seeds known as bunya nuts. Heavy crops occur about every 3 years. Aboriginal people gathered in the Bunya Mountains for bunya festivals, during which they harvested the cones' protein-rich seeds.[24]

Why Endangered: Although bunya nuts are highly prized as a food source, bunya forests have been heavily logged, so the tree populations remain low.[25]

Efforts to Save: Although Bunya Mountains National Park was established in 1908, logging continued illegally for many years. Increased ecotourism and enforcement of protections have helped maintain the remaining bunyan forests.[26]

Sorghum, millets, Kenya

Special Characteristics: These grains, which are micronutrient-dense, plant-based sources of protein, are widely grown, hardy, drought-resistant crops.[27]

Why Endangered: Sorghum and millets are seen as good sources of nutritious food for the impoverished. But as maize, used for free food aid, and rice have become more popular, local people have lost the knowledge of how to cook sorghum and millet.[28]

Efforts to Save: ICRISAT and Africa Harvest are working with rural communities to grow, cook, and develop the value chain of millet and sorghum, in part through agribusiness projects with women in local communities.[29]

Guinea hog, United States

Special Characteristics: At one time, this was the most popular pig breed in the U.S. Southeast. The hog foraged for its own food; ate rats, snakes, and other pests; and cleaned out garden beds. It was an inexpensive, easy-to-raise source of tender, high-protein ham, bacon, and lard for subsistence farmers.[30]

Why Endangered: The pig is highly prized by butchers and pastry chefs (its lard is used in crusts and dough). As Southern homestead habitat disappeared over time, the hog became extremely rare.[31]

Efforts to Save: There is growing interest from U.S. farmers in raising rare livestock breeds as alternatives to factory-farmed animals. The American Livestock Breeds Conservancy is using a $150,000 grant to encourage and educate people on the benefits of conservation and more sustainable rearing methods.[32]

Holland chicken, United States

Special Characteristics: These chickens were developed as a dual-purpose breed that would lay white eggs (preferred and more profitable in mid-1900s). They are well suited to farm conditions, good foragers, and cold tolerant, and they raise their own offspring. The breed produces medium to large protein-rich eggs.[33]

Why Endangered: They are highly sought-after because their slow to moderate growth rate limits mass production. Their yellow skin is preferred by American consumers, and they can produce a large quantity of eggs over their lifetimes.[34]

Efforts to Save: This is one of the preferred species for free-range eggs, which encourages greater attempts to conserve the species through this method of rearing.[35]

Abouriou grapes, France

Special Characteristics: This grape grows quickly and produces high yields while also being disease resistant. It is useful for blending and is used to make Malbec, Cabernet Sauvignon, Cabernet Franc, Merlot, Syrah, Fer, and Gamay wines.[36]

Why Endangered: The climate is shifting where the grapes grow, and production is done by cooperative wineries, using very intrusive techniques, such as machine harvesting and selected yeasts. Almost half the Abouriou vineyards were pulled up between 2007 and 2011.[37]

Efforts to Save: There is growing interest in this grape variety, with California vineyards planting vines.[38]

In many ways, Mukiibi is fighting the continued spread of agricultural practices that focus on yields above all else. To make matters worse, prices for commodity crops are often depressed because so many farmers are growing them.[39]

Farmers in India face similar challenges, and in a perverse twist the results there are often deadly. Thousands of poor Indian farmers have committed suicide since the introduction of Green Revolution technologies in the country some 60 years ago. Many of these farmers were encouraged to purchase—often on credit—expensive inputs and seeds. When the crops failed to produce high enough yields or the prices were too low to pay back their debt, they resorted to suicide.[40]

But many other farmers, including India's Khasi tribe, are finding ways to avoid this cycle of debt, poor yields, and poverty. Instead of consuming diets of just a few crops, such as maize or rice, the Khasi are still doing what they have been doing for centuries: saving seeds and growing diverse food crops. Approaches like this, says Daniel Moss, executive director of the AgroEcology Fund, "are hedges against too much or too little rain—trends likely to worsen as climate change turns weather topsy-turvy. Their practices reduce risks of malnutrition and market uncertainties."[41]

Another organization, the North East Slow Food and Agrobiodiversity Society, is helping farmers who grow indigenous crops find markets for locally and regionally grown fresh and healthful foods. For example, the farmers are selling organic millet, rice, potatoes, and vegetables to residents in Shillong, an Indian city of about 500,000 people. According to Moss, the farmers are working to ensure the quality of their crops through what are called Participatory Guarantee System groups, which allow them to certify producers "based on active participation of stakeholders and are built on a foundation of trust, social networks, and knowledge exchange."[42]

Certification helps give credibility to indigenous foods—traditional foods are often considered poor people's foods or even weeds—which don't get the respect they deserve from the scientific or agricultural funding communities. But that is slowly changing as well.

Research groups such as the World Vegetable Center work with farmers to maintain and breed vegetable varieties that serve their specific needs. The center is based in Taiwan but has regional offices throughout sub-Saharan Africa, Asia, Central America, and Oceania.[43] Since its founding, the center "has distributed more than 600,000 seed samples to researchers in the public and private sectors in at least 180 countries."[44] It also maintains a gene bank of the world's largest public vegetable germplasm collection, with more than 61,000 accessions from 155 countries. This includes roughly 12,000 accessions of indigenous vegetables, helping preserve the crops farmers have been breeding on their own for future generations.[45]

In addition, the International Center for Research in the Semi-Arid Tropics (ICRISAT) has developed the Smart Food Initiative to create demand for an often-forgotten group of cereal: the millets, which ICRISAT calls the "super crops of our ancestors."[46]

Millets are small-seeded grasses, widely grown around the world for consumption by people and livestock. Millets tend to be hardy, resilient crops that can withstand changes in weather, and more than 90 million people in Africa and Asia depend on them as a staple part of their diets.

Millets, according to ICRISAT, "are often the only cereal crops that can grow in arid lands, needing only 350–400 mm annual rain,"[47] so their value will only increase as rainfall patterns change in the face of climate change.

Millets also tend to be high in iron, have a low glycemic index, are gluten free, and are rich in other important vitamins and nutrients such as zinc and calcium. One strain, finger millet, has three times more calcium than milk from cows.[48]

ICRISAT's Smart Food initiative hopes to do a couple of things. First, it wants to build a stronger scientific and funding case for supporting millets. Second, it wants to create consumer demand for these nutrient-rich grains. Although millets have been traditionally eaten as a porridge, they can also be used in a variety of value-added products and help increase incomes for the farmers who grow them.[49]

It's projects like these and dozens of others that will help increase recognition of, demand for, and increased research and investment in the often-forgotten foods that make the global food system so diverse.

Preserving these foods may be the best way to preserve cultural traditions and mores that would otherwise be lost. In much of the world, feeding yourself, and your family, involves more than choosing foods at the market—it entails rationality, tradition, memory, symbols, and values based on religion, race, and gender. All of these are quite literally wrapped up in what we choose to grow and eat.

Controlling Food: Food and Power Roles

Access to food and nutrition—what might be called the nutritional order and hierarchy—is a function of power. Those with the most power tend to eat first and more compared with those with less power.[50]

Historically, the control of food has been one of the principal sources of power over others. In the Middle Ages, the banquets of aristocratic families contrasted with near starvation among the peasants. There was little recourse for the less powerful, and in various parts of Europe, those who were caught poaching in the royal reserves or in the preserves of local lords were often sentenced to death. Countless battles and wars have been waged between farmers and stockbreeders in many regions of the world, not only over access to land but also over control of the techniques and tools used for producing and preserving food. In parts of Africa, these types of conflicts are still going on.[51]

Food can signify power in social terms, especially regarding social prestige. But the cultural perception of that accepted meaning is fairly complex and in some cases contradictory. Certain foods have always been recognized in terms of prestige and exclusiveness, especially for reasons linked to high cost and insufficient availability. Foods such as chocolate, which is now available to nearly everyone, were once accessible only to the wealthiest people. Meat typically was consumed in large quantities only by the rich, while among the poor it was used more like a condiment, to add flavor and protein. But anthropologists and sociologists have found that the gradual spread of well-being and technology has caused a progressive tapering of the divide in food preferences and habits between the rich and the poor.[52]

However, it's not only food that signifies power but also land. The phenomenon of land grabbing has occurred for centuries. But today, because of climate change, migration, and other issues, the stakes are much higher. The land grabs occurring today are taking valuable land and natural resources from farmers, particularly in sub-Saharan Africa and South America. Land grabs are defined by the International Land Coalition as land "deals that happen without the free, prior, and informed consent of communities that often result in farmers being forced from their homes and families left hungry." Both countries, including China and Saudi Arabia, and corporations have been involved in these deals, which can lead to the displacement of farming communities and the loss of food culture, as well as human rights abuses and other social problems.

Box 4.1. Rethinking the Meat We Eat

In 1971, author and food justice advocate Frances Moore Lappé wrote *Diet for a Small Planet.* The book detailed, with the statistics then available, why a meat-centered diet was bad for people and the planet. For most Americans, this was borderline heresy: At that time meat was the center of the plate, and calling its role into question was almost un-American. But Frances Moore, as she was then known, made a compelling case for a diet rich in vegetables, grains, beans, and other vegetarian protein. She was also one of the first food advocates to assert that eating and our daily food choices could be political.[a]

In particular, as she described more than 40 years ago, too much of the world's grain was being fed to animals, not people. The growth of industrial animal operations, or factory farms, across the globe has made this situation even more troubling today.

Lappé commented on the grain problem, saying, "I understand, of course, that grain-fed meat is not the cause of the world hunger problem—and eating some of it doesn't directly take food out of the mouths of starving people—but it is, to me, a symbol and a symptom of the basic irrationality of a food system that's divorced from human needs. Therefore, using less meat can be an important way to take responsibility. Making conscious choices about what we eat, based on what the Earth can sustain and what our bodies need, can help remind us that our whole society must begin to balance sustainable production with human need."[b]

However, meat continues to be a highly desired food. In Western countries, especially those with Anglo-Saxon origins, meat is esteemed from a nutritional standpoint because it is rich in protein and other nutrients that are crucial to a balanced diet. In the Global South, meat is also highly regarded, although in many places that's largely because it is not eaten regularly. In this part of the world, meat and other animal products are typically eaten by those with higher incomes or during special events, including religious and cultural celebrations.

At the same time, meat can inspire fear and worry. Food scares and scandals have been a part of American history since Upton Sinclair detailed the practices of the livestock and canned meat industries in *The Jungle*, which was

a. A. Aubrey, "If You Think Eating Is a Political Act, Say Thanks to Frances Moore Lappé," National Public Radio (2016), www.npr.org/sections/thesalt/2016/09/22/494984095/70s-food-movement-promoted-benefits-of-plant-based-diet

b. F.M. Lappé, *Diet for a Small Planet* (New York: Ballantine, 1971).

published in 1906. And even today, issues such as bovine spongiform encephalopathy, or mad cow disease, as well as other zoonotic diseases, such as severe acute respiratory syndrome or meat contaminated with *E. coli*, create distrust. Even the World Health Organization has warned eaters about the dire health effects—including some kinds of cancer—from eating processed meats.

Religious taboos also inspire fear around different kinds of meat. For example, Muslims ban pork, Hindus ban beef, and Jews have special rules about mixing meat and dairy and eschew eating shellfish.

Even in the present day, eating meat continues to play the role of social aggregator. That said, recent food scandals and new medical discoveries regarding excessive meat consumption[c,d] have led people to change their diets to reduce their meat consumption, as per World Health Organization recommendations.[e]

What's more, being a vegetarian is no longer seen as something strange or weird. Even in a place like Italy, which has deep culinary traditions around meat, there has been a consistent increase in the consumption of soy-based foods. Soy products are familiar to four out of five Italians and are served in as much as 40 percent of Italian households. According to data from Eurispes, the percentage of Italians who have stopped eating meat went from 4.9 percent in 2013 to 5.7 percent in 2015. A recent study promoted by the dairy cheese cooperative TreValli in collaboration with Eurisko highlights how ethical and health issues have motivated more than 2 million Italians to reduce their meat consumption and another million to give up animal-based products, including honey, entirely. In the United States, some 3.2 percent of adults, or 7.3 million people, call themselves vegetarian.[f]

---

c. S.C. Larsson and N. Orsini, "Red Meat and Processed Meat Consumption and All-Cause Mortality: A Meta-Analysis," *American Journal of Epidemiology* 179, no. 3 (2014): 282–9, https://doi.org/10.1093/aje/kwt261

d. S. Rohrmann et al., "Meat Consumption and Mortality: Results from the European Prospective Investigation into Cancer and Nutrition," *BMC Medicine* 11 (2013): 63, https://www.ncbi.nlm.nih.gov/pmc/articles/PMC3599112/

e. WHO and FAO, "Diet, Nutrition and the Prevention of Chronic Diseases," World Health Organization Technical Report Series no. 916 (2002), http://www.who.int/dietphysicalactivity/publications/trs916/en/

f. "Vegetarians and Vegans: Feeding the Future?," *L'Eurispes.it* (2015), www.leurispes.it/vegetariani-vegani-alimentazione-futuro/; "Italians, It Changes: The Board Becomes Green," *Trevalli Cooperlat* (2015), www.trevalli.cooperlat.it/news-502-italiani-si-cambia-la-tavola-diventa-verde; "Vegetarianism in America," *Vegetarian Times, Clean Eating* (2017), https://www.vegetariantimes.com/uncategorized/vegetarianism-in-america

There is new thinking and advocacy about the role of meat—especially meat from factory farms—in our diets. Campaigns such as Meatless Mondays encourage consumers to forgo meat and other animal products at least 1 day a week, and prominent chefs such as Dan Barber and Jose Andres have put vegetables at the center of many of their dishes, creating cauliflower and carrot "steaks" or enhancing the flavor and texture of vegetables, beans, and grains in such a way that their customers often don't even notice that meat is not part of the entrée. Producers such as Paul Willis of Niman Ranch are, somewhat counterintuitively, encouraging consumers to eat less meat overall. The company's customers tend to carefully consider where their meat comes from and how it's raised, and they're willing to spend more money on meats that meet their high standards. This attitude is also becoming more mainstream; Niman Ranch now supplies humanely raised pork from pasture-raised pigs to Chipotle, the enormous Mexican food chain. According to "The Power of Meat," a report written by the Food Marketing Institute and the North American Meat Institute, year-on-year sales of organic meat and poultry grew 32 percent in the United States in 2015, to a 12-month total of $569 million in November. There has also been a 29 percent increase in the volume of organic meat and poultry sold over the same period, to 92 million pounds.[g]

g. "The Power of Meat 2016 Report," Food Marketing Institute and North American Meat Institute (2017), https://goaptaris.com/wp-content/uploads/2016/06/Power-of-Meat-2016.pdf

In the past 10 years, more than 81 million acres around the world have been sold to foreign investors. The crops grown on these lands are typically cash or commodity crops, such as sugarcane, palm oil, and soy, and 60 percent are exported instead of used by local communities, two-thirds of which are facing food scarcity problems.[53]

Women, in particular, feel the impact of traditional power structures that tend to be discriminatory. Sociologist Marjorie DeVault highlights how the female practice of preparing food in the home, however gratifying it might be, subtly reveals the pervasive conditions of inequality and subordination. This inequality is often mirrored in agricultural communities themselves.[54]

Women make up about 43 percent of the agricultural labor force worldwide, and in some countries they make up 80 percent of all farmers, according to the FAO. In addition to tending crops, most women, particularly in the Global South, are also responsible for seed saving, animal husbandry, grain processing, and other tasks related to growing food as well as for cooking, cleaning, and taking care of sick elders and children.

Furthermore, it's women farmers who typically produce the food that families eat. While male farmers often focus on growing commodity crops such as maize, rice, and soybeans, women raise the fruits, vegetables, and small livestock that nourish families from day to day.

At the same time, these women have little agency over their own lives. They often lack the same access to resources, including land, banking and financial services, education, and extension services, as male farmers. In many countries, women aren't allowed to own land or even inherit it, putting them, and their land, at risk when their husbands die.

But around the globe, women farmers are letting governments, policymakers, and their own husbands, brothers, fathers, and sons know that we ignore women in the food system at our own peril. According to the FAO, if women had the same access to resources as men, they could raise current yields by 20 to 30 percent and lift as many as 200 million people out of hunger.[55]

For example, a group of women farmers in Niger, working with ICRISAT, established a communal garden several years ago that uses solar drip irrigation to grow vegetables, fruit trees, and other crops. They eat part of what they grow and also sell produce, ornamentals, and trees in Niamey, the country's capital. Before they started the garden, these women had been earning about $300 a year apiece, or less than a dollar a day. Today, they're making about $1,500 a year. Still, it's more than money these women have gained. They're innovating their way to a sustainable life and becoming an example to their children.

In Ghana, a small group of women dairy farmers, working with the nongovernment organization Heifer International, started a small dairy cooperative to make yogurt to sell to local schools and stores. At first their husbands were angry that the women dared to start the group without their permission. But as they saw their family incomes grow and saw how the women were using the money—to pay for healthcare for their children and to send them to school—their anger turned to respect.

Thousands of miles away, in the United States, an organization called FarmHer is on a mission to highlight the contributions of female farmers. The group was founded to change the face of agriculture, so to speak, in eaters' minds, and to remind all of us that women have been and always will be an important part of small-, medium-, and large-scale farming.

These women and organizations are changing the food system and making it more sustainable for all of us.

Traditional power structures often ignore the role of youth in agriculture as well, which is particularly perilous when you consider that farmers around the globe are aging. In the United States, according to the latest agricultural census, the average farmer is 57.3 years old; in sub-Saharan Africa, surprisingly, the average age is roughly the same, although for different political and social reasons. For the next generation of agricultural leaders to feel empowered—whether they're farmers, food entrepreneurs, or scientists—they need access to education, mentoring, and financing that help them do their jobs better.

The Barilla Center for Food & Nutrition (BCFN) Young Earth Solutions contest encourages youth to put big ideas into action. Every year, the contest challenges young PhD students and researchers to come up with actionable ideas to make the food system more sustainable. The winners are given scholarships of 20,000 (about US$23,500) to continue their work. The 2016 winners, Shaneica Lester and Anne-Teresa Birthwright of Jamaica, won for their project centered around curriculum for small farmers on various climate-adaptive irrigation strategies. The young women hold workshops with local farmers about techniques they've developed to adapt to and mitigate the effects of climate change. This kind of work not only helps Lester and Birthwright put their ideas into practice while still in school, it also brings farmers into the research process.

Young farmers and researchers also need mentorship, including learning how to advocate for themselves. The Young Professionals in Agricultural Research organization, housed at the Global Forum for Agricultural Research in Rome, brings together researchers under 40 and gives them the chance to network not just with one another but with seasoned agricultural professionals. In the United States, the National Young Farmers Coalition (NYFC) helps young farmers mobilize, engage, and direct more attention to issues facing them, including lack of access to land, mentorship, and education about the business of farming as well as how to deal with more general challenges, such as student loan debt and healthcare. When asked to name the biggest opportunity for changing the food system, the group's executive director, Lindsey Shute, says, "The biggest opportunity lies in the talent and ambition of young farmers. If they're given a real chance of success—land to own, sufficient capital, healthcare, and appropriate technical support—they will thrive and change the food system through their entrepreneurship."[56]

Creating a culture of equality can be a key part of preserving agriculture for generations to come, ensuring that women, youth, and others have access to the resources they need to do their jobs as food producers, entrepreneurs, researchers, scientists, and educators.

The Great Culinary Tradition of the Mediterranean Diet and the Reality of Food Today

There is a deep-seated link between food and culture. On one hand, food has a significant effect on people's lives, and on the other hand, ways of eating reflect and are conditioned by the habits and the relationships people create. In some parts of the world, the interaction between these variables has given rise over time to unique and very specific dietary approaches and gastronomical traditions.[57]

The Mediterranean style of eating, which developed as a result of geography and climate, is now considered not only a reflection of the Mediterranean culture but one of the healthiest ways of eating in the world.[58]

The term "Mediterranean diet" was developed by Ancel Keys, a doctor and nutritionist at the University of Minnesota Department of Food Science and Nutrition, who later wrote the book *Eat Well and Stay Well.* He spent time in Italy and noticed that that the less well-to-do, or poor, in Italy ate a diet of mainly bread, onions, tomatoes, and vegetables but were healthier than urbanites in the United States. Keys developed the Seven Countries Study, which compared diets of more than 12,000 people ranging in age from 40 to 59 in Finland, Greece, Italy, Japan, The Netherlands, the United States, and the former Yugoslavia. The study found an association between diet and the incidence of chronic noncommunicable disease. People who eat a Mediterranean diet have a lower incidence of heart attack compared to countries such as Finland where the diet is rich in saturated fats, including butter, lard, dairy products, and red meat. The final result of the Seven Countries Study indicated that the best dietary regimen was followed by inhabitants of Nicotera in southern Italy, where they ate a very simple yet diverse Mediterranean diet of olive oil, bread, pasta, garlic, red onions, aromatic herbs, vegetables, and very little meat. Compared with the other nations in the Seven Countries Study, Italy, and specifically Nicotera, had very low levels of blood cholesterol and low incidence of heart disease.[59]

Beginning in the Neolithic Age, the Mediterranean Sea was the destination of countless migrations. The new arrivals settled in existing communities in search of better living conditions: more fertile soil for those who came from Asian or African deserts, for example, or a milder climate for those coming from Scandinavia or Germany. In the 11th and 12th centuries, the interaction between Muslim and Christian communities, based on the Iberian Peninsula, grew into commercial exchanges during which a significant number of new food products were introduced into the respective gastronomical cultures.[60]

In the high Middle Ages, the ancient Roman tradition—which, on the model of Greek culture, identified bread, wine, and olive oil as the products symbolizing the tradition of a farming and agricultural civilization—encountered the culture of the Germanic peoples. Those peoples lived in forests, where they hunted wild game, herded livestock, foraged mushrooms and other foods, and grew crops. These interactions created a new food culture.[61]

It was the Muslims, however, who gave rise to a significant process of agricultural renewal, in which irrigated fields played a fundamental role. This new agriculture method introduced unfamiliar plant varieties that, because of their high prices, had been used only by the more prosperous social classes. Products introduced into Mediterranean cuisine that originally came from the Islamic world include sugarcane, rice, citrus fruit, eggplant, spinach, and spices. Rose water, orange water, lemon water, almond water, and pomegranate water were introduced as well. These foods helped change the cultural unity of the Mediterranean, which had been constructed by force on the Roman model, providing a decisive contribution to the new gastronomic tradition that was taking shape.[62]

Another event with historic impact was the conquest of America by the Europeans. This led to the exchange and trade of other food products, such as potatoes, tomatoes, corn, peppers, chili peppers, and a number of varieties of beans. Not that these exchanges were always seamless. For instance, Europeans initially thought of the tomato as no more than an exotic curiosity and an ornamental fruit, and it wasn't until much later that they considered it edible. Now, of course, it has become a symbol of Mediterranean and, in particular, Italian cuisine. Although the central role of vegetables is one of the most distinctive characteristics of the Mediterranean tradition, it is also important to remember the role played by cereal grains, as the foundation of poor people's cuisine and as a means of day-to-day survival. Grains fill people up, especially the poor, and can be stored for longer than fruits and vegetables, making them an important staple in the Mediterranean diet.[63]

The food model that is now called the Mediterranean diet is not only a way of nourishing oneself but also the expression of an entire cultural system that's based on healthfulness, the quality of food, conviviality, and a love of food. In fact, commensality, or dining with others, has always played a central role in the social lives of Mediterranean people. In a literal sense, this term means "eating at the same table"; it comes from the Medieval Latin word *commensalis*, from *con-dividere* (to share) and *mensa* (table). In a broader sense, the word conveys the idea of habitually sharing food and sometimes implies the dependency that dining companions may have on one another. The Italian term *partecipare* ("to participate"), for example, comes from the Latin *pars capere*, which literally means "receiving your own portion of a sacrificial meal"; this participation implies that you belong to and have your own place in a group, an institution, or an event.[64]

Unfortunately, the Mediterranean diet and culture, like so many other food traditions, faces threats. Even the eating habits of Italians are changing, and many of the country's young people are abandoning the Mediterranean diet for other foods, particularly fast foods, that are high in fat and sodium. According to a 2016 study by the BCFN, some 24 percent of the roughly 105 million meals eaten every day in Italy are eaten outside the home, and Italians increasingly prefer to eat lunch out (53 percent) over dinner out (47 percent). Italians' fast-paced lifestyles and long work hours are affecting how they eat. About 9 percent of Italians surveyed said they often eat lunch on the go and in less than 10 minutes, and 14 percent eat while walking from one place to another. This is similar to food and dining trends in most other countries in Europe.[65]

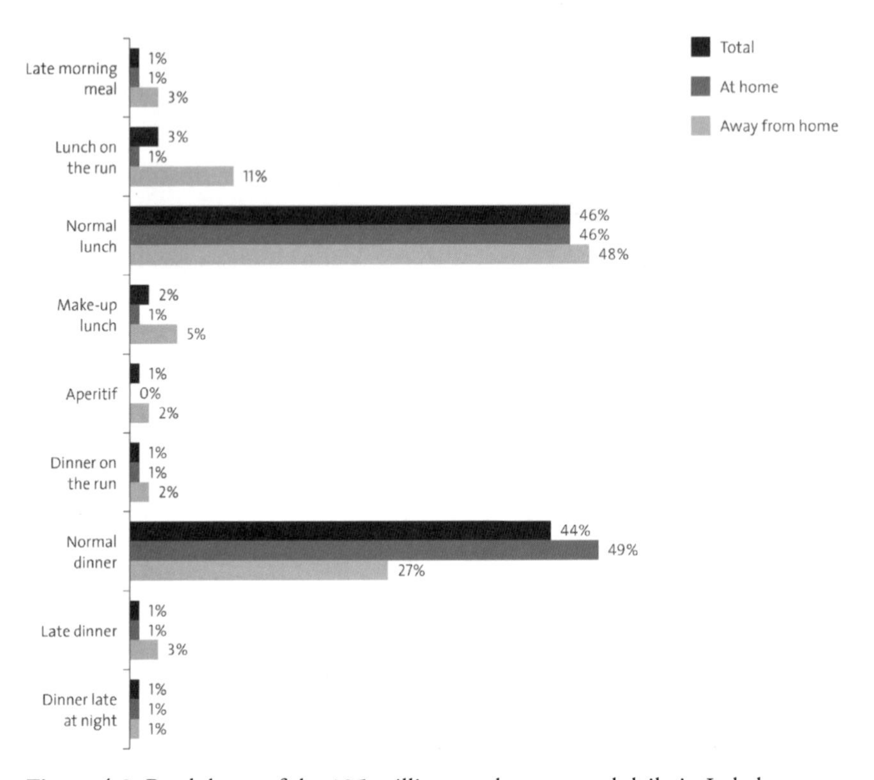

Figure 4.3. Breakdown of the 105 million meals consumed daily in Italy by mode of consumption. Note: Data expressed in %. Base: 99,000 meals analyzed; 105 million meals daily. Source: BCFN on Nielsen-Barilla data, 2009, from p. 253 of *Eating Planet*, 2nd ed. (Milan: Edizioni Ambiente and the Barilla Center for Food & Nutrition, 2016).

And although Italians have historically had the longest life expectancy and the lowest weight in Europe, that, too, is changing. Today, nearly 2 in every 10 Italian teenagers are overweight, and the country has one of the highest rates of childhood obesity in Europe (9.3% of Italian children are obese). The situation is not much better for adults and the elderly.[66]

In addition, fewer and fewer young people and adults are not physically active; in fact, only 3 in 10 people in Italy exercise regularly.

The combination of these of two factors—sedentary lifestyles and a change to diets rich in animal proteins and fats—can lead to increased rates of illnesses such as diabetes, cardiovascular disease (which remains the main cause of death around the world), and other chronic diseases (which cause 60 percent of deaths globally).[67]

According to sociologist Claude Fischler, the Mediterranean approach to food now has become unexpectedly fragile. It is unexpected not just because its health benefits are well known but because in the past it was the Mediterranean diet, more than any other, that had proved capable of successfully assimilating elements of other cultures without losing its distinctive personality. It had a secure, clear, and strong identity, but today that identity is being lost as other food ways, particularly the growth of fast and so-called convenience foods, gain prominence. Not only are these foods often much lower in nutritional content, they also lack sociability and significance. Unfortunately, the Mediterranean diet does not have a strong mechanism for transmitting tradition. This puts the people of the Mediterranean at risk of losing a wealth of knowledge and nutritional behaviors that exist nowhere else.[68]

Still, the Mediterranean diet is well respected by scientists and food experts across the globe as one of the healthiest diets in existence. A report titled "Can We Say What Diet Is Best for Health?," by food expert and nutritionist Dr. David Katz, points out that the Mediterranean diet results in a positive ratio of omega-6 to omega-3 essential fatty acids, a high-fiber diet, and a diet rich in antioxidants and polyphenols. The study also supported findings that link the Mediterranean diet with "increased longevity, preserved cognition, and reduced risk of cardiovascular disease in particular, with some evidence for reduced cancer risk."[69]

Professor Antonino De Lorenzo, director of the School of Specialization in Food Science at the University of Rome Tor Vergata and an expert in Mediterranean diet and nutrition, commented on the reputable scientific data that promote the popularity and efficacy of the diet in living a healthy life. He stated, "The Mediterranean Diet is the most famous diet in the world, declared an Intangible Cultural Heritage Site by UNESCO, and universally recognized as the optimal regimen for healthy eating."[70]

Although the trend away from the Mediterranean diet raises tremendous concerns in Italy and the rest of the world, it can be reversed. Reestablishing home economics classes in primary and secondary schools or after-school programs can help students learn basic cooking skills—skills they can take home to their parents. For example, Tony Hillery, founder of the project Harlem Grown in New York City, teaches kids to grow and harvest vegetables and fruits that were previously unfamiliar to them, including tomatoes and broccoli, and he teaches them about the importance of healthy, nutritious diets and cooking what they harvest. And chefs such as Jamie Oliver are working in schools and using cooking shows aimed at children to help them learn about nutrition.[71]

The Mediterranean diet's focus on eating diverse and colorful foods is also beneficial. The colors of fruits and vegetables are signs of nutritional content. A richly colored red tomato has high levels of carotenoids such as lycopene, which the American Cancer Society reports can help prevent cancer and heart disease. The correlation between nutrients and color is true for other foods, too. Eggs with bright-orange yolks, for instance, are high in cancer-fighting carotenoids and are more likely to be produced by healthier chickens that are not raised in industrial operations.[72]

There also may be a link between happiness and the consumption of fruits and vegetables. According to Dr. Andrew Oswald of Warwick University in the United Kingdom, eating more fruits and vegetables can improve a person's mental well-being, separate from other variables such as income level and how much meat that person eats. This research is supported by a similar study from the Harvard School of Public Health that found a link between patients' blood levels of carotenoids, compounds commonly found in colorful fruits and vegetables, and their feelings of optimism.[73]

The Mediterranean diet also focuses on seasonality, stressing that when foods are in season, they're not only at their peak flavor, they also tend to retain more nutrients, because they haven't traveled long distances. Eating seasonally can also inspire a sense of greater community, whether by buying a share directly from a farmer participating in a community-supported agriculture program, shopping at farmers' markets, or even growing some of your own food in a community garden.[74]

Conviviality and sharing meals—Mediterranean or not—can also increase feelings of well-being. Talking and laughing while sharing food is a uniquely human experience. Conviviality and joyful, friendly interaction are found at markets and around dinner tables, and these shared experiences can support healthy relationships and healthy bodies. The BCFN considers convivial food culture one of the most critical aspects of food and agriculture, alongside health, hunger alleviation, and sustainable development. Researchers from Cornell University and the University of Minnesota agree, reporting that the benefits of family dinners for children's mental health and achievement levels depend on engagement with their parents at these meals.[75]

What's more, the Mediterranean diet can be good for the environment. The Double Pyramid model (see Chapter 2), based on the Mediterranean diet, describes how eating fewer animal products and more fruits and vegetables can be a way to simultaneously maintain a healthy weight, improve nutrition, and protect the planet. And this ancient way of eating—again, consuming a variety of fruits and vegetables, nuts, nutritious grains, and fish, and avoiding red meat, high-fat foods, and sodium—might very well be the diet of the future.[76]

Conclusion and Action Plan

In ancient times, societies tended to organize their experiences with food around just a few things—activity, or work, specifically, and free time. But today, our interactions with food are very different. Most of us no longer grow our own food; many of us don't even cook it ourselves.[77]

A variety of forces determine what and how we eat, how much we eat, where we eat, and what we can get access to and afford, and often these forces compete against one another. For example, although globalization and the homogenization of cultures are leading to the loss of foods, consumers, especially young consumers, are increasingly interested in local and regional foods and concerned about their health and transparency in the food system.[78] Millennial eaters in the United States, in particular, also desire a connection to their food and the story behind it, wanting a connection to the experience of eating.[79] At the same time, Western consumers still expect their food to be inexpensive. In fact, they spent 4 to 10 percent of their total budgets on food in 2014, whereas people in other countries sometimes spend more than 50 percent of their incomes on food.[80]

What is needed is a new a vision for food and culture that goes forward by going back and recognizing the rich traditions and flavors that have nourished people for centuries and combining them with new research and innovation around food that help people eat and live better, whether they live in Brooklyn or Botswana.

The food of the future certainly needs to be tasty, but it also needs to be produced sustainably. Grains such as kernza, a perennial crop that has deep roots and needs little water, can be grown in harsh climates. It's also versatile and can be used to make everything from bread to beer.[81] In Africa, raising livestock on grasslands not only provides protein and income but, if done sustainably, can also help restore degraded lands and improve biodiversity, according to research from the Zimbabwe-based Savory Institute. Rediscovering forgotten foods and so-called orphan crops is also more important than ever: Foods such as jackfruit and breadfruit, which are even being used as a replacement for meat, are growing in popularity in Africa and Asia.[82]

We also need to find ways to make all aspects of the food system—from farming and processing to research and policymaking—more attractive career options for young people. Agriculture can be "cool" if youth see that it can help them earn a decent income and provide the intellectual stimulation and variety they crave in their working lives. Creating better internet and cellular infrastructure in sub-Saharan Africa and other rural and agriculture-based economies can help keep young, creative people on farms by giving them access to the resources they need. Edie Mukiibi's work at Slow Food International not only helps create awareness around forgotten foods, it also helps connect farmers to agronomists and extension agents they can call on with questions or concerns about their crops and livestock. A connected farmer can send a simple text or leave a message with agronomists, and the response time is short, usually less than a few days.

As we look ahead, it's also increasingly clear that we must democratize the food system. Farmers need to be involved in all aspects of research and development, and well-meaning development agencies and funders need to find ways to be more inclusive and listen to communities' needs and wants first.

And there can no longer be a system of haves and have-nots. Good food is not only for the wealthy. Instead, nutritious, healthy, and sustainable foods must be accessible and affordable for everyone. We can no longer afford an agricultural system that values quantity over quality; we need to do more than fill people up with empty calories and starchy staple crops. We have to nourish both their minds and bodies. In India, for example, the Self-Employed Women's Association (SEWA), one of the largest unions in the world, grows organic lentils, rice, and other foods that the women brand and market under the SEWA label. Other SEWA members then sell these foods to city dwellers, providing a source of high-quality but low-cost food to the urban poor.[83]

The funding and donor communities, in particular, need to change their strategies around everything from food aid to their policies on fighting climate change. We know today that simply donating U.S. grain to other countries is not the best strategy for fighting hunger. Although these shipments may provide much-needed calories, they disrupt local markets, lowering prices for locally grown foods and driving out producers. A better long-term strategy is the local and regional procurement of food aid and allowing neighboring nations to help one another when food aid is needed. This not only provides an important market for farmers but can also teach them skills to help them be part of the global market.[84]

Moreover, we need to recover the concept of conviviality and sharing meals. Recently, cooking and sharing meals within a community or school setting has gained traction and popularity. In eco-communities, such as the CoHo Ecovillage in Corvallis, Oregon, residents inhabit small plots of land, rely on communal resources, and cook and share meals together in a co-housing community center. This practice decreases food and water waste and energy use; it is also more economical, associates mealtime with increased social interactions, and reinforces the community's cultural values.[85]

For another example of meal sharing, consider the supper clubs organized by the Society of Grownups, which hosts gatherings that are simultaneously meals and financial investment classes for millennials. By combining food sharing with financial education, millennials nourish themselves in multiple ways.[86]

The World Food Programme also offers meal sharing in an educational setting, and as of 2015 the group provided school meals to 17.4 million children in 62 countries across the globe. Its program provides nutritious food to students who do not receive proper nutrition at home, incentivizing parents to send their children to school and improving educational achievement, because students who are not hungry are better able to focus on their schoolwork.[87]

The future of the food system will shift from focusing primarily on what we eat to include how we eat, creating the opportunity to adopt healthful eating habits, improve food quality, and provide more access and affordability. Eaters need to rethink their relationship with food and how it supports health, social justice, and sustainability. In other words, we need to learn from the old adage: Tell me how you eat and what you eat, and I'll tell you who you are.

To create real food system change

- The funding and donor communities need to invest more research into the benefits of indigenous and traditional foods for economic and environmental sustainability.
- Farmers and producers need to rethink the types of meat and other animal products we raise and eat, focusing on breeds that are hardy and resistant to drought and disease.
- Consumers need to work on finding ways to refine their consumption of animal products, focusing on quality, health, sustainability, and humane treatment.
- Governments need to take a stand against foreign acquisition of agricultural land, or land grabbing, which can destroy agricultural communities and leave farmers destitute.
- Policymakers and community leaders and elders need to value the work of women, not only as farmers and food producers but as nutritional gatekeepers, caretakers of rich agricultural traditions, and stewards of the land and biodiversity.
- Eaters, farmers, nutritionists, scientists, and the funding and donor communities need to recognize the value of the Mediterranean diet for both the environment and human health.
- We need to value the role of farmers, not only as food producers but as businesswomen and businessmen, entrepreneurs and innovators, and stewards of the land who should be recognized for the ecosystem services they provide that benefit us all.

VOICES FROM THE NEW FOOD MOVEMENT: Natasha Bowens

Can you talk about why you decided to interview other farmers of color in the United States? Why was it important to you to explore their stories?

When I started immersing myself in the food movement, I was working on healthcare and also really passionate about environmental and social justice issues. A light went off, and I saw that food was at the center of everything that I was working on. I jumped into the food movement. I wanted to grow my own food and learn how to do it sustainably. I also wanted to address the racial and economic inequities that I was seeing within the system. I really fell in love with food and agriculture. As a woman of color myself, I was instantly shocked and frustrated at the lack of diversity in the movement. Everywhere I was going—conferences, farmers' markets, even the books I was reading—it just seemed to represent a very exclusive movement. At the same time, people of color were the ones that seemed to be impacted most heavily by this same broken food system. It just didn't add up, and I thought, "Why aren't we hearing from farmers of color?" After doing some research, I realized that the only way to hear their stories would be to hit the road and go hear them for myself. It became a very personal journey. As a biracial woman, I struggled with my new identity as a beginning farmer because I felt like this movement that I was joining wasn't welcoming for people of color.

In your introduction, you talked about how during your involvement with the good food movement in different parts of the United States, you noticed a discernible absence of people of color. Why was it important for you to trace the intersections of food

and race, and as you mention, "redefine our agrarian identity as people of color"?

Once I started asking questions and writing about the intersection of race and agriculture on my blog, Brown.Girl.Farming., which helped me express my thoughts while I was out there farming, the research that I was doing and the things that I was finding taught me that my agrarian story was already written. Most of what I could find that discussed the intersection of race and agriculture was about our past, and it's a very negative past of slavery, migrant farm labor, and oppression of people of color in the fields. I felt that this narrative, too, was very exclusive; it was this very pigeonholed agrarian story for people of color. I knew that a different narrative was out there for my people, one inspired by a legacy of living on the land and deep knowledge and wisdom. So, I really wanted to redefine that while still raising awareness about the negative aspects of our history that we have to address. But I didn't want that mainstream history to be the only thing that defined my agrarian story.

Your book provides many examples of resilience and persistence among farmers of color across the United States, despite the many obstacles stacked against them. What were some of the most difficult barriers for farmers of color, and how have they dealt with it?

The book starts out with the section "Rooted in Rights," which talks about water and land rights, and specifically land ownership and land loss. African-American people recognize and acknowledge that we were once owned as property, and that very shortly after our emancipation, black land ownership skyrocketed in a way that was kind of shocking. As descendants of the enslaved, when you think about what it would be like to have once been property and then suddenly progressed to owning your own property and being able to grow food and sustain your family, it's kind of stunning. To me, that is the epitome of resilience, and that story continues today as we keep struggling with land rights and land loss and land ownership, as well as with issues like accessing resources, accessing markets, and all of the discrimination that is deeply woven still into many aspects of our food system. Around the time that I hit the road to write *The Color of Food*, the *Pigford v. Glickman* case was coming to a close. That was the case filed by black farmers against the USDA for racial discrimination that drove many black farmers into foreclosure and loss of their farms because they were not afforded the same access to resources and loans as white farmers.

One of the prevalent themes in your book, throughout its many poignant portraits of farmers of color, is access to land. You just discussed the history of land loss in the United States. How did farmers interviewed for *The Color of Food* respond to these challenges in their local communities?

Some people ask me what my biggest surprise was getting out on the road. I thought I would find a lot of farmers that were angry and still trying to fight the system, maybe getting involved on the political level to change things so that they could have an equal footing in agriculture. Instead, what I found were really revolutionary farmers who were completely bucking the system and figuring out their own ways to access capital or the things they needed to run their farms. They were going back to concepts their ancestors invented, like cooperative farms, communal farming, and cooperative marketing for crops in order to overcome a lot of these barriers. It seems to me that the number one goal of a lot of these folks was to preserve and grow their communities, something they even prioritized above having successful businesses, though they had those, too. A lot of these farmers were bucking the system and focusing on their communities. Food that was relevant to that community and the preservation of culture, tradition, storytelling, and heirloom-seed centers were special to these communities. It was this revolutionary way of saying, "There are barriers here, but we are going to operate outside the system as much as we can to continue resiliently moving forward."

How does the revival of food and farming traditions empower communities of color?

I think it's invaluable and empowering because it circled back to my experience of getting into farming and that negative stigma for people of color. The history of farming suggests that we are always farming for someone else, building someone else's business, profit, and power while never being able to build your own independence and power. Even today, we're often growing for others and operating on these big agribusiness farms. Often farmworkers live in "food deserts," where they are growing food but are left hungry. It's the complete opposite of empowering, whereas things like what Kevin Welch is doing are really reviving and facilitating the remembrance of tradition and agricultural techniques. We're really bringing those back for ourselves, not for someone else. To build our own profit, business, community, and, therefore, power. To me, this reclamation is the most empowering thing that we can be doing.

How are the women farmers that you interviewed for *The Color of Food* spearheading the way forward for small family farmers and sustainable agriculture through their sustainable practices and innovation?

I knew that I wanted to dedicate an entire section to the fierce farming women out there. Similar to farmers of color, I found that a lot of the women were farming for their communities. They got into farming for their families and communities because they were sick and tired of the lack of food choices. Or, like in Nelida's case, a farmer in Washington State, she was sick and tired of the corporate-controlled agricultural system in which she worked as a migrant laborer. So she decided that she wanted to start her own farm and run it organically. I felt like the reasons that a lot of women got into farming were far more empowering and far more community-minded than the average farmers. I also found that a lot of the women farmers were spearheading a lot of the cooperatives and community efforts, like community kitchens and community markets. Even though a lot of the farmers I met were operating outside of the system, when I did meet farmers who were heavily involved in it as food activists and advocates for food policy councils, for example, they were, more often than not, women. There was Jenga Mwendo down in New Orleans, Louisiana, who was creating a food policy council there to change the shape of the food movement in the Lower Ninth Ward. I felt like the women were far more revolutionary and fiercer, hence the name for that section.

Can you describe the importance of having young farmers of color involved in the good food movement and some of the main lessons learned from the young farmers you interviewed for *The Color of Food*?
As we know, the agricultural industry is aging out rapidly. The average age of the farmer is about 57, but for black farmers, for example, the average is 63. As far as young farmers entering the field, our numbers as young farmers of color are much lower than young white farmers. It is absolutely vital that we are picking up the pitchfork today and joining the movement, and it doesn't just have to be farming. There are a lot of food activists and farm educators that I interviewed who were playing a lot of different roles in the movement. It comes down to the central message of the book: If we want to have a truly sustainable food system that is inclusive and taps into the answers that our ancestors and our overlooked communities know to be right, then our leaders have to be from these communities. If we want to have land ownership, or food sovereignty over our food from seed to table, then we have to be the ones picking up the pitchfork. We can't have an exclusive movement representing us. We have to stand up and be heard as people of color. We have to represent ourselves.

We young, new, and beginning farmers of color face all the same barriers that older farmers of color face, but one of our biggest barriers is land access and capital. One of the young farmers that taught me the most—a couple actually—were Cristina and Tahz of Tierra Negra Farms. While we young farmers are supposed to be picking up that pitchfork and carrying it forward, we can't do it alone, and we can't do it without the guidance of our elders and those who came before us. That was one of the biggest lessons that Cristina and Tahz had to teach me. They were surrounding themselves with what they called "eldership." Their region is home to the Occaneechi Band of the Saponi Nation, a Native American tribe in the Piedmont region of the Carolinas. They were reaching out and surrounding themselves with the native elder community there, who were teaching them a new, communal way of looking at land ownership.

How do you define food sovereignty? How would you say the farmers you talked to are working toward it?

Food sovereignty is all-encompassing and all-empowering. To me, it means full ownership over our food, from seed to table. If we don't want genetically modified seeds, then we have to start saving and owning our own seeds and starting seed collectives. Seed banks are an example of what I see some of these farmers doing to advance their food sovereignty. Moving forward, we need to find ways to grow our own food, own our own land, and have ownership over our right to water. Southwestern farmers in New Mexico and Arizona are using the ancient *acequia* system to achieve these goals, a democratic ownership system which makes water a common. Instead of dividing up the food movement to have people working on farmworker rights, people working on local and organic food, people working on food justice, we should transition to food sovereignty, which is the all-encompassing movement-builder in my mind. We have to have ownership and fair wages for everyone working in the food system, whether they are picking, processing, or selling our food. A lot of the indigenous farmers who I interviewed taught me that food sovereignty comes down to what we eat and how we eat it. We've got this movement towards nutrition and healthy food, with education programs in communities of color with widespread health issues. But this movement is trying to introduce foods that aren't native to these communities. That is also an integral part of food sovereignty: being able to have ownership over the decision of what to eat. We should be able to say, "This is what my people eat."

Natasha Bowens is a farmer, community activist, and the author of *The Color of Food: Stories of Race, Resilience and Farming*. Her book explores the intersections between food and race as she tells the stories of black, Latino, Asian, and indigenous farmers and food activists in North America through photography, oral history, and storytelling. As the organic and local food movement has taken shape in the United States over the last two decades, Bowens has worked to bring visibility to the issues faced by farmers of color in a way that celebrates culture, tradition, and community.

Citation: "Food Sovereignty and Farmers of Color: An Interview with Natasha Bowens," Food Tank (2016), https://foodtank.com/news/2015/08/food-sovereignty-and-farmers-of-color-an-interview-with-natasha-bowens/. *Interview conducted by Claudia Urdanivia in August 2015 and edited by Michael Peñuelas in August 2017.*

VOICES FROM THE NEW FOOD MOVEMENT: Lindsey Shute

What originally inspired you to get involved in your work?

I started organizing with young farmers because of the challenges that my husband Ben and I faced in growing our own farm. As we met more and more farmers who were facing similar struggles across the nation, I realized that we lacked a political voice. There were too many young people with the ambition and will to farm, but without a way to get there.

What makes you continue to want to be involved in this kind of work?

I have a constant source of inspiration and motivation in the people whom I work to represent: young farmers. These farmers are out to change the country by growing great food, taking care of the soils and water that they depend on, and daring to compete as small farmers in a highly consolidated food system. The risk that these farmers take on behalf of their communities keeps me going. I want them to succeed, and I know what they're up against.

I'm also encouraged by our bi-partisan traction and success at cutting through partisan divides. Just a few weeks ago, Rep. Glen Thompson (R–PA) and Rep. Joe Courtney (D–CT) reintroduced the Young Farmer Success Act (H.R. 1060) to add farmers to the Public Service Loan Forgiveness Program. These co-sponsors were joined by two additional Republicans and two additional Democrats. These actions demonstrate how farming can be unifying—and a way to overcome national divisions in favor of help for ordinary people.

Who inspired you as a kid?

My two grandfathers were rural ministers and World War II chaplains. As a child, they served as beacons of service, faith, and devotion to community that I can only hope to achieve. When I would attend my family's church in southeast Ohio as a child, the day would be filled with stories from church members about how my grandfather made a difference in their lives. One particular story that stuck with me is about a neighbor boy who repeatedly robbed my grandfather's farmhouse. Over the course of months, electronics went missing and eventually my grandfather's gun. After the kid went to jail on other counts, my grandfather repeatedly visited him and expressed his forgiveness and hope for the kid's future.

What do you see as the biggest opportunity to fix the food system?
The biggest opportunity lies in the talent and ambition of young farmers. If they're given a real chance of success—land to own, sufficient capital, healthcare, and appropriate technical support—they will thrive and change the food system through their entrepreneurship. The candidate for Secretary of Agriculture, Sonny Purdue, can leverage this new talent by directing the USDA to stand by young people in agriculture.

Can you share a story about a food hero who inspired you?
Leah Penniman is one of my food heroes. Last spring, I gave a short, public talk about why we have lost so many farmers in the United States and I failed to speak to the effects of racism. Leah, in the audience at the time, rightly let our team know that my narrative was incomplete. Her willingness to speak up in that moment and to continue dialogue with our team led to the development of a racial equity program at NYFC—as well as more farmers of color identifying with and joining the coalition. Leah helped me in that moment, and I am deeply grateful for her strong voice and leadership.

What's the most pressing issue in food and agriculture that you'd like to see solved?
With the massive cuts proposed at USDA, healthcare access for farmers on the brink of collapse, and immigration enforcement threatening the

farm workforce, it's hard to ignore the myriad of rural issues created by the Trump Administration. But outside of these immediate policy crises, the nation must address the issue of affordable land access for farmers. In the next 20 years, two-thirds of the farmland in the United States will change hands as our aging farm population retires. How that land transitions will set the stage for the future of our food system. If we provide access for working-class, small farmers, we will promote economic vitality, national security, and sustainability.

What is one small change every person can make in their daily lives to make a big difference?

Practice empathy. In so many of the political discussions that I've been hearing recently, there has been so much antipathy for people facing struggle. We critique immigrants who, like most of us, came here for good work and opportunity. We call out folks who couldn't afford healthcare before the Affordable Healthcare Act, and we undermine government programs that stoke innovation in areas of the country where mobility and economic agency have grown dim. I believe we need smart government programs that leverage best practices in technology and management, but I also want a government that stands by the principles of empathy and compassion for our neighbors. To get there, we need empathy. And to practice empathy, we probably need to drop our phones and make time for conversation with people outside our immediate circles.

What advice can you give to President Trump and the U.S. Congress on food and agriculture?

Agriculture is the wealth of the nation, and a large part of our national security. Although so few Americans are now farming, these farmers have an outsized impact on the nation's health and prosperity. We need to invest in their futures and ensure that we are supporting smaller farms that minimize risk, make our economy more resilient, and keep dollars in rural communities.

Lindsey Shute is the executive director and co-founder of the U.S.-based NYFC. With a background in organizing and state policy, Shute co-founded NYFC as a platform for young, progressive farmers to have a meaningful influence on the structural obstacles in the way of their success. In addition to her work with NYFC, Lindsey is co-owner of Hearty Roots Community Farm in Tivoli, New York.

Citation: "Lindsey Shute, National Young Farmers Coalition: 'We need empathy,'" Food Tank (2017), https://foodtank.com/news/2017/03/lindsey-shute-interview/. *Interview edited by Michael Peñuelas in August 2017.*

VOICES FROM THE NEW FOOD MOVEMENT: Stephen Ritz

What inspired you to get involved in food and agriculture?
Food is a non-negotiable. It is a point of entry for everyone and a daily need/input that all humans share in common. In a community with limited means and limited access to healthy food, the notion of embedding food and agriculture into public education and workforce development skills seemed like an absolute add value proposition to me. While I am growing food and vegetables, my vegetables are growing students, citizens, school performance, and communities. The entire movement is about planting seeds—both literally and figuratively. Seeds represent genetic potential, and my goal is to make sure all my students and my colleagues reach and fulfill their God-given genetic potential; crops well planted and tended can give you a harvest of abundance and epic proportion.

What do you see as the biggest opportunity to fix the food system?
The biggest opportunity to fix the food system comes via education. First and foremost, we need to teach children what food is and what food isn't. We need to teach them dignity and respect for their bodies—that what we put into them determines what comes out of them. We need to teach children to respect the earth, the inputs, and embedded energy that go into food production. We need to teach children to respect farmers—who remarkably, and more often than not, share the same cultural backgrounds. We need to educate children about media and marketing, to move them from being consumers to producers and changing their status in the larger ecosystems and economic systems that they are a part of. When children grow food in school, they learn all of this and

become actively empowered and engaged in their lives and the lives of living things around them.

What innovations in agriculture and the food system are you most excited about?

I'm phenomenally inspired by the entire good food movement—from active policy makers to those on the front lines working with farmers and people daily. The urban agriculture movement, vertical farming movement, indoor controlled agriculture movement—be it hydroponics, aeroponics, aquaculture, permaculture—are tremendous opportunities to redefine the work we do to feed the planet and care for it as well. The movement towards transparency, understanding sourcing, labeling, and fair and responsible labor practices gives me hope for a better, brighter, more equitable future for all.

Can you share a story about a food hero that inspired you?

There are so many heroes in this movement—those in public view and those laboring in far off lands far under the radar. That said, I want to shout out children who are becoming the voice of the movement, who are demanding and working for environmental and social justice and are holding themselves and their leaders accountable. For this movement to move beyond sustainability and become regenerative for people and the planet, we need active and empowered youth—individuals committed to fight the good fight and leave the world better than they found it.

What drives you every day to fight for the bettering of our food system?

Simply put, I am an equity warrior. I believe in people and the planet with a triple bottom line orientation. For so many, food is the problem, yet for all of us, including mother Earth, responsible practices can offer solutions! Business as usual is no longer an option. For me, this all starts with children—it is easier to raise healthy children than fix broken men!

What's the biggest problem within the food system our parents and grandparents didn't have to deal with?

One of the biggest problems I see today is marketing, corporate spending, and the way big business targets our youth. Children are inundated with messaging, logos, subliminal messages, and shameless marketing ploys via celebrities almost every waking moment of the day, on so many platforms; we as responsible adults and parents owe them better. The proliferation of cheap food, fast food, and convenience food has forever changed the landscape of the world and the waistlines of our populations—far too often on the hearts, backs, minds, and lifespans of the marginalized and poor.

What's the first, most pressing issue you'd like to see solved within the food system?

Waste—it is appalling—across every sector and segment of society and business. It is also tied to the marketing issue and the notion of a picture-perfect world or picture-perfect food. We need to stop perpetuating lies and creating false images. To think how much food goes to waste daily sickens me.

What is one small change every person can make in their daily lives to make a big difference?

Simply put, eat less meat; the impact on your body and the planet will be phenomenal. Eat as locally—in every sense of the word—and as seasonally as possible. I'm a big advocate of the notion: act locally, think globally! Hippocrates said it best: "Let thy food be thy medicine, and let thy medicine be thy food." Imagine if we embraced that wisdom on a daily basis!

What's one issue within the food system you'd like to see completely solved for the next generation?
Hunger! We have an abundance on the planet—we need to address this issue immediately. Nobody should go to sleep hungry. This also ties into my concerns with waste and diversion. We also need to end the bucket brigade mentality of solutions and look at systems thinking! On the flip side, it is appalling to realize that obesity has now become the face of hunger for so many.

What agricultural issue would you like for the next president of the United States to immediately address?
I am an educator first and foremost, but I am very excited to see the movement towards increasing minimum wage and farm worker conditions. Treating workers, farmers, and the land with dignity and respect seems like a fine agenda to address. Dignity, transparency, and respect matter!

Stephen Ritz is a lifelong educator and founder of the Green Bronx Machine, a kindergarten-to-12th-grade food production module integrated into curricula. The organization also transformed an underused library into a state-of-the-art facility including a year-round commercial indoor vertical farm and food processing and training kitchen with renewable energy generation.

Citation: "Ten Questions with Stephen Ritz, Founder of the Green Bronx Machine," Food Tank (2016), https://foodtank.com/news/2016/04/ten-questions-with-stephen-ritz-founder-of-the-green-bronx-machine/. *Interview edited by Michael Peñuelas in August 2017.*

VOICES FROM THE NEW FOOD MOVEMENT: Ruth Oniang'o

What or who has inspired your work?

It all started from my childhood because I saw my parents feed people in our community who were unable to feed themselves. They knew that if they came by our gate with empty plates, my mother would have food waiting. Now, I go around doing what I do, and people say, "You take after your parents!"

What are your professional goals for the future?

My professional goals are mainly to consolidate what I've already been doing. I have been speaking nationally and internationally in forums about the need to support small-holder farmers, especially women in sub-Saharan Africa. My goals also include advocating for those who are tied to small-holder farmers. It has been so difficult to get big donors to trust or build the capacity of NGOs who are directly tied to small-holder farmers. I also hear a lot of talk from them about gender issues, and specifically on supporting women and girls, but I do not see the resources being brought to bear.

What changes do you feel are necessary for the food system today?

Changes have been needed for a long time at various levels. Firstly, we need the advocacy to continue. We have to be persistent. Persistence at whatever level can realize change. The other changes we need to make are on the donor side. They need to support agriculture and food security. You cannot have hungry, malnourished people supporting their own continent. Donors need to make a paradigm shift in how they look at us so that they better understand what the issues are. We also need to work to attract young people, both men and women, into agriculture in order to help national and continental developments. At the same time, we need a proper policy environment to encourage local and international investors. Finally, on that note, we need the private sector to create platforms where you can address the food value chain from production to consumption. You need everybody at the table in order to realize change.

What has been your greatest obstacle? How have you overcome it?
Initially, when I started working on the ground, it was near the area where I was married, and then where I was born. I came out of there, and I want to see others, especially girls, coming out of there. As time went on, I was able to start seeing women stand up in meetings where it was both men and women. Men in the community now support women to take up jobs as chiefs in the village. And now, here in Kenya, we have many women in the Parliament. Breaking that barrier was a huge challenge—making sure women get the respect that they deserve.

What is your greatest success?
Professionally, I have had various successes. I was a member of the Kenyan Parliament, and we were able to pass legislation to professionalize nutrition in this country. I think Kenya is doing much better than other countries on that front. In terms of the knowledge of nutrition, I can proudly say that I have been at the forefront. I was always part of the Kenyan team for technical support, helping to prepare the country papers. I am the Chair of the Sasakawa Africa Association, co-established by late Japanese philanthropist Ryoichi Sasakawa, late scientist and Nobel Peace Prize Laureate Dr. Norman Borlaug, and the former President of the United States Jimmy Carter in 1986. When I speak internationally and give examples of African countries, it is because I have actually been there on the ground. I also run the *African Journal of Food, Agriculture, Nutrition and Development* (AJFAND). It is reaching all corners of the world. It hosts conversations about food security and nutrition, with a focus on Africa. I started

AJFAND in 2001 as a capacity-building facility to assist budding African scientists to share their scholarly work and to give them visibility.

What advice do you have for young women hoping to make a difference in the food system today?

My idea to encourage young people is to make them aware that there is so much space along the value chain to make a difference. It can be virtually anything! But at the end of the day, we are all concerned about how the world will be fed with an increasing population. My appeal to young people is that it doesn't matter what you're doing, there is always room for you to be concerned about how the world eats, because we all eat.

Why do you feel that supporting local economies and farmers is so critical in your work?

It bothers me that our leaders do not care about the hungry. When children are hungry, they cannot learn, and without an education, we cannot develop as a nation or as a continent. I feel that Africa, as a continent, should compete with other continents, and that Kenya, as a country, is not poor. There is no reason for us not to develop the same as other continents. I graduated with distinction from college. I got a scholarship to do both my bachelor's and master's degrees. We can compete. That is what I keep telling young people, young women.

Why do you feel that it's important to support female farmers and women in agriculture?

Today we have a situation where women are on farms, but nobody is advising them. I am just a food scientist who felt that women needed their voices heard and decided that I was going to be that voice. I've known that women, worldwide, are very valuable, but in this rural community, they are extremely vulnerable because of the patriarchal structure of our society. If a woman is ejected from her husband's home, she cannot go anywhere else. I am coming from a community that has so many issues with this. At the end of the day, you just hope you can do what you can and make an impact in somebody's life.

Ruth Oniang'o is a professor, author, and former member of parliament in Kenya's National Assembly. She has taught at multiple universities and is the founder and director of the Rural Outreach Programme, a nonprofit organization that works to empower rural communities to mobilize their existing resources and strengths. Oniang'o is also founder and editor-in-chief of the *African Journal of Food, Agriculture, Nutrition and Development* (AJFAND), an internationally recognized publication that covers topics including agriculture, food, nutrition, environmental management, and sustainable development.

Citation: "Leading Globally, Inspiring Locally: An Interview with Food Scientist and Professor Ruth Oniang'o," Food Tank (2014), https://foodtank.com/news/2014/11/leading-globally-inspiring-locally-an-interview-with-food-scientist-an/. *Interview conducted by Cordi Craig in 2014 and edited by Michael Peñuelas in August 2017.*

Notes

Chapter 1

1. Field visit to Zasaka, Danielle Nierenberg, June 2015.
2. Prolinnova, Promoting Local Innovation in Ecologically Oriented Agriculture and Natural Resource Management, www.prolinnova.net/, viewed 10 August 2017; Self Employed Women's Association (SEWA), www.sewa.org, viewed 10 August 2017; field visit to Ahmenabad, India, Danielle Nierenberg, November 2011.
3. "Threats: Soil Erosion and Degradation," World Wildlife Fund (2017), https://www.worldwildlife.org/threats/soil-erosion-and-degradation
4. "Earth Day," Barilla Center for Food & Nutrition (2017), https://www.barillacfn.com/en/media/earth-day-22nd-april-2017-the-bcfn-foundation-and-the-milan-center-for-food-law-and-policy-each-year-the-planet-loses-an-agricultural-surface-area-as-big-as-the-philippines-in-40-years-30-of-all-arable-land-will-be-infertile-while-hunger-is-already-a-risk/
5. C. DeLong, R. Cruse, and J. Wiener, "The Soil Degradation Paradox: Compromising Our Resources When We Need Them the Most," *Sustainability* 7 (2015): 866–79.
6. R. Cruse, "Iowa State Project Aims to Reduce Major Cause of Soil Erosion on Iowa Farm Fields," Iowa State University (2014), https://www.cals.iastate.edu/news/releases/iowa-state-project-aims-reduce-major-cause-soil-erosion-iowa-farm-fields
7. Ibid.
8. DeLong et al., "The Soil Degradation Paradox."
9. Ibid.
10. E. Rusco, B. Marechael, M. Tiberi, C. Bernacconi, G. Ciabocco, P. Ricci, and E. Spurio, "Sustainable Agriculture and Soil Conservation Case Study

Report: Italy," JRC European Commission and the Institute for Environment and Sustainability (2008).

11. Ibid.
12. DeLong et al., "The Soil Degradation Paradox."
13. S. Scherr, "Soil Degradation: A Threat to Developing-Country Food Security by 2020?" International Food Policy Research Institute (1999).
14. DeLong et al., "The Soil Degradation Paradox."
15. Ibid.
16. V. Bado, "Farmers' Perception of Land Degradation and Solutions to Restoring Soil Fertility in Niger," International Crops Research Institute for the Semi-Arid Tropics (2017), www.icrisat.org/farmers-perception-of-land-degradation-and-solutions-to-restoring-soil-fertility-in-niger/
17. X. Diao and D.B. Sarpong, "Cost Implications of Agricultural Land Degradation in Ghana: An Economy-Wide, Multimarket Model Assessment," International Food Policy Research Institute (2007), http://www.ifpri.org/publication/cost-implications-agricultural-land-degradation-ghana
18. Scherr, "Soil Degradation."
19. Ibid.
20. Ibid.
21. R. Kumar, "World Day to Combat Desertification: ICRISAT'S Definitive Steps Towards a Greener Africa," International Crops Research Institute for the Semi-Arid Tropics (2017), www.icrisat.org/world-day-to-combat-desertification-icrisats-definitive-steps-towards-a-greener-africa/
22. Scherr, "Soil Degradation."
23. Ibid.
24. F. Santibanez and L. Berry, "In Coquimbo Region–Chile: Its Extent and Impact," GM and the World Bank (2003).
25. Ibid.
26. E. Barbier, "Carbon Farming: A Solution to Global Land Degradation and Poverty?," The Conversation (2012), www.theconversation.com/carbon-farming-a-solution-to-global-land-degradation-and-poverty-8491
27. Ibid.
28. S. Small, "It's All About Soil," Food Tank (2015), https://foodtank.com/news/2015/01/twenty-fifteen-declared-the-international-year-of-soils/
29. J. Glover, "How to Create Resiliency in Food and Agriculture," Food Tank Summit at George Washington University (2017), https://foodtank.com/

news/2017/03/create-resiliency-food-agriculture-food-tank-summit-2017-gwu/

30. "Biography: Jerry Glover, Agroecologist," *National Geographic* (2017), http://www.nationalgeographic.com/explorers/bios/jerry-glover/
31. S. Snapp, "Perennial Grains: Transformative Option or Pipe Dream?," *Contested Agronomy Conference Proceedings* (2016), https://www.youtube.com/watch?v=s1SwE1Dd1zc
32. S. Snapp et al., "Biodiversity Can Support a Greener Revolution in Africa," *Proceedings of the National Academy of Sciences* 107, no. 48 (2010), http://www.pnas.org/content/107/48/20840.short
33. M.A. Florentín et al., "Green Manure/Cover Crops and Crop Rotation in Conservation Agriculture on Small Farms," UN Food and Agriculture Organization, *Integrated Crop Management* 12 (2010), http://www.fao.org/fileadmin/user_upload/agp/icm12.pdf
34. Snapp et al., "Biodiversity Can Support a Greener Revolution in Africa."
35. É. Maillard and D.A. Angers, "Animal Manure Application and Soil Organic Carbon Stocks: A Meta-Analysis," *Global Change Biology* 20, no. 2 (2014): 666–79, https://www.ncbi.nlm.nih.gov/pubmed/24132954; L. Xia, S.K. Lam, X. Yan, and D. Chen, "How Does Recycling of Livestock Manure in Agroecosystems Affect Crop Productivity, Reactive Nitrogen Losses, and Soil Carbon Balance?," *Environmental Science and Technology* 51, no. 13 (2017): 7450–7457, http://pubs.acs.org/doi/abs/10.1021/acs.est.6b06470
36. P. Shrestha, "Prolinnova Global Partnership Program: Relevance for Asia-Pacific," Expert Consultation to Review Progress of Agricultural Research Networks and Consortia in Asia-Pacific (2007), www.prolinnova.net/sites/default/files/documents/resources/presentations/apaari-prolinnova_presentation_oct07.pdf
37. J. Barrett, N. Cherret, and R. Birch, "Exploring the Application of the Ecological Footprint to Sustainable Consumption Policy," Stockholm Environment Institute, New York, NY; L.K. Williams, "Challenges, Opportunities for Water Reuse in the Food & Beverage Industry," *Industrial Water World* (2017), www.waterworld.com/articles/iww/print/volume-12/issue-04/feature-editorial/challenges-opportunities-for-water-reuse-in-the-food-beverage-industry.html

38. NASA/Goddard Space Flight Center, "Summer Sea Ice Melt in the Arctic," *ScienceDaily* (24 July 2017), https://www.sciencedaily.com/releases/2017/07/170724133153.htm
39. "Climate Change: How Do We Know?," NASA (2017), https://climate.nasa.gov/evidence/; W. Legg and H. Huang, "Climate Change and Agriculture," *OECD Observer* 278 (2010), www.oecdobserver.org/news/archivestory.php/aid/3213/Climate_change_and_agriculture.html; S. Vermeulen, B. Campbell, and J. Ingram, "Climate Change and Food Systems," *Annual Review of Environment and Resources* 37 (2012), http://www.annualreviews.org/doi/full/10.1146/annurev-environ-020411-130608
40. "Double Pyramid: Healthy Food for People, Sustainable Food for the Planet," BCFN (2016), https://www.barillacfn.com/m/publications/pp-double-pyramid-healthy-diet-for-people-sustainable-for-the-planet.pdf
41. Ibid.
42. G. Eshel, A. Shepon, T. Makov, and R. Milo, "Land, Irrigation Water, Greenhouse Gas, and Reactive Nitrogen Burdens of Meat, Eggs, and Dairy Production in the United States," *Proceedings of the National Academy of Sciences of the United States of America* 111, no. 33 (2014): 11996–2001, www.pnas.org/content/111/33/11996.full
43. F. Burchi and P. De Muro, "From Food Availability to Nutritional Capabilities: Advancing Food Security Analysis," *Food Policy* (2015), www.sciencedirect.com/science/article/pii/S0306919215000354; "Food Security: Policy Brief," Food and Agriculture Organization (FAO) of the United Nations (2006), http://www.fao.org/forestry/13128-0e6f36f27e0091055bec28ebe830f46b3.pdf
44. "Starvation Stalks Nigeria, Somalia, South Sudan, Yemen," Aljazeera (2017), www.aljazeera.com/news/2017/04/starvation-stalks-nigeria-somalia-south-sudan-yemen-170411101432521.html
45. "The Cost of the Double Burden of Malnutrition," World Food Programme and the Economic Commission for Latin America and the Caribbean (2017), www.es.wfp.org/sites/default/files/es/file/english_brochure_april_26_2017.pdf
46. "Globally Almost 870 Million Chronically Undernourished—New Hunger Report," Food and Agriculture Organization (FAO) of the United Nations (2012), www.fao.org/news/story/en/item/161819/icode/

47. "The State of Food Insecurity in the World 2015," Food and Agriculture Organization (FAO) of the United Nations (2015), www.fao.org/hunger/glossary/en/; A. Shattuck and E. Holt-Gimenez, "Why the Lugar–Casey Global Food Security Act Will Fail to Curb Hunger," Food First (2009), https://foodfirst.org/publication/why-the-lugar-casey-global-food-security-act-will-fail-to-curb-hunger/; L.A. Thrupp, "Linking Agricultural Biodiversity and Food Security: The Valuable Role of Sustainable Agriculture," *Royal Institute of International Affairs* 76 (2000): 265–81; S. Martinez, M. Hand, M. Da Pra, S. Pollack, K. Ralston, T. Smith, S. Vogel, S. Clark, L. Lohr, S. Low, and C. Newman, "Local Food Systems: Concepts, Impacts, and Issues," United States Department of Agriculture Economic Research Service (USDA ERS) (2010), https://www.ers.usda.gov/webdocs/publications/46393/7054_err97_1_.pdf?v=42265
48. "The American Food Paradox: Growing Obese and Going Hungry," *Working Knowledge: Business Research for Business Leaders*, Harvard Business School (18 December 2017), www.hbswk.hbs.edu/item/the-american-food-paradox-growing-obese-and-going-hungry
49. Ibid.
50. L. Dinour, D. Bergen, and M.C. Yeh, "The Food Insecurity–Obesity Paradox: A Review of the Literature and the Role Food Stamps May Play," *Journal of the American Dietetic Association* 107, no. 11 (2007): 1952–61.
51. "Poverty," The World Bank (2016), www.worldbank.org/en/topic/poverty/overview; "The World's Cities in 2016," United Nations, Department of Economic and Social Affairs, Population Division (2016), http://www.un.org/en/development/desa/population/publications/pdf/urbanization/the_worlds_cities_in_2016_data_booklet.pdf
52. "Growth and Poverty Reduction in Agriculture's Three Worlds: World Development Report 2008," The World Bank (2008), https://openknowledge.worldbank.org/bitstream/handle/10986/5990/9780821368077_ch01.pdf
53. E. Dabla-Norris, K. Kochhar, N. Suphaphiphat, F. Ricka, and E. Tsounta, "Causes and Consequences of Income Inequality: A Global Perspective," International Monetary Fund (2015), https://www.imf.org/external/pubs/ft/sdn/2015/sdn1513.pdf
54. M. Visciaveo and F. Rosa, "Volatilità dei prezzi agricoli: un confronto fra prodotti e paesi dell'Ue," *Agriregionieuropa* 8, no. 31 (2012).

55. "International Year of Family Farming: Successes of Cover Cropping Explored in National Survey," Food Tank (2013), https://foodtank.com/news/2013/09/successes-of-cover-cropping-explored-in-national-survey/; "Using Agroforestry to Save the Planet," Food Tank (2013), https://foodtank.com/news/2016/05/using-agroforestry-to-save-the-planet/; "The Regeneration Hub: Mapping the Regeneration Movement," Food Tank (2017), https://foodtank.com/news/2017/03/regeneration-hub-mapping-regeneration-movement/; "Reduce Greenhouse Gas Emissions by Eating Less Meat, CGIAR Report Says," Food Tank (2016), https://foodtank.com/news/2016/10/reduce-greenhouse-gas-emissions-by-eating-less-meat-cgiar-report-says/; "Sustainable Agriculture and Climate Change," BCFN (2011), www.barillacfn.com/wp-content/uploads/2012/11/pp_agricoltura_sostenibile_cambiamento_climatico_eng.pdf
56. P. Sundell and M. Shane, "The 2008–09 Recession and Recovery: Implications for the Growth and Financial Health of US Agriculture," United States Department of Agriculture (2012), https://www.ers.usda.gov/publications/pub-details/?pubid=40484
57. R. Naylor and W. Falcon, "Food Security in an Era of Economic Volatility," *Population and Development Review* 36, no. 4 (2010): 693–723, http://citeseerx.ist.psu.edu/viewdoc/download?doi=10.1.1.472.5979&rep=rep1&type=pdf
58. M. Ivanic and W. Martin, "Implications of Higher Global Food Prices for Poverty in Low-Income Countries," The World Bank Policy Research Working Paper Series (2008), https://ssrn.com/abstract=1149097
59. "How High Food Prices Affect the World's Poor," World Food Programme (2012), https://www.wfp.org/stories/how-high-food-prices-affect-worlds-poor
60. "Eating Planet Food and Sustainability: Building Our Future," BCFN (2016), https://www.barillacfn.com/media/pdf/Barilla_Eating_planet-2016_Eng_S_abstract.pdf
61. H. Kharas, "Making Sense of Food Price Volatility," Brookings (2011), https://www.brookings.edu/opinions/making-sense-of-food-price-volatility/
62. D. Nierenberg, "Hedging against Hunger: Our Best Bet to Avoid a Global Food Bubble," *Huffington Post* (3 February 2013); R. Bailey, T. Benton, et al., "Extreme Weather and Resilience of the Global Food System: Final

Project Report from the UK–US Taskforce on Extreme Weather and Global Food System Resilience," The Global Food Security Programme, UK (2015), http://ilsi.org/publication/extreme-weather-and-resilience-of-the-global-food-system/; G. Tadesse, B. Algieri, M. Kalkuhl, and J. von Braun, "Drivers and Triggers of International Food Price Spikes and Volatility," *Food Policy* 47:117–128 (2014), https://www.sciencedirect.com/science/article/pii/S0306919213001188

63. D. Nierenberg and K. Paramaguru, "Betting on Hunger: Is Financial Speculation to Blame for High Food Prices?," *Time* (12 December 2012), http://science.time.com/2012/12/17/betting-on-hunger-is-financial-speculation-to-blame-for-high-food-prices/?iid=sr-link7; R. Sharma, "The Next Global Crash: Why You Should Fear the Commodities Bubble," *The Atlantic* (16 April 2012), https://www.theatlantic.com/business/archive/2012/04/the-next-global-crash-why-you-should-fear-the-commodities-bubble/255901/; "Excessive Speculation in Agricultural Commodities," Institute for Agriculture and Trade Policy (IATP) (2011), http://www20.iadb.org/intal/catalogo/PE/2011/08247.pdf; J. Vidal, "UN Warns of Looming Worldside Food Crisis in 2013," *The Guardian* (13 October 2012), https://www.theguardian.com/global-development/2012/oct/14/un-global-food-crisis-warning
64. D. Nierenberg and F. Kaufman, "The Food Bubble: How Wall Street Starved Millions and Got Away with It," *Harper's Magazine* (July 2010), https://frederickkaufman.typepad.com/files/the-food-bubble-pdf.pdf
65. D. Nierenberg, "Oxfam Gets Behind Our Demands for Regulation of Food Speculation," *Global Justice Now* (6 October 2011), http://www.globaljustice.org.uk/blog/2011/oct/6/oxfam-gets-behind-our-demands-regulation-food-speculation; B. Scott "A Guide to Food Speculation: How to Argue with a Banker," *The Ecologist* (9 June 2011), https://theecologist.org/2011/jun/09/guide-food-speculation-how-argue-banker
66. D. Nierenberg, T.A. Wise, and S. Murphy, "Resolving the Food Crisis: Assessing Global Policy Reforms Since 2007," Institute for Agriculture and Trade Policy (IATP) (2012), http://www.ase.tufts.edu/gdae/Pubs/rp/ResolvingFoodCrisis.pdf
67. J. Gleaves, "Bet the Farm: Six Questions for Frederick Kaufman," *Harper's Magazine* (2012), https://harpers.org/blog/2012/10/bet-the-farm-six-questions-for-frederick-kaufman/

68. I. Perez, "Climate Change and Rising Food Prices Heightened Arab Spring," *Scientific American* (2013), https://www.scientificamerican.com/article/climate-change-and-rising-food-prices-heightened-arab-spring/; C. Werrell and F. Femia, "The Arab Spring and Climate Change: A Climate and Security Correlations Series," Center for American Progress, Stimson, and the Center for Climate and Security (2013), https://climateandsecurity.files.wordpress.com/2012/04/climatechangearabspring-ccs-cap-stimson.pdf
69. "Severe Droughts Drive Food Prices Higher, Threatening the Poor," The World Bank (2012), www.worldbank.org/en/news/press-release/2012/08/30/severe-droughts-drive-food-prices-higher-threatening-poor; "Food Price Watch," The World Bank and Poverty Reduction and Equity Group (2012), www.siteresources.worldbank.org/EXTPOVERTY/Resources/336991-1311966520397/Food-Price-Watch-August-2012.pdf
70. "Eating Planet Food and Sustainability: Building Our Future," BCFN (2016), https://www.barillacfn.com/media/pdf/Barilla_Eating_planet-2016_Eng_S_abstract.pdf
71. S. Goldenberg, "Climate Change 'Already Affecting Food Supply'—UN," *The Guardian* (2014), https://www.theguardian.com/environment/2014/mar/31/climate-change-food-supply-un; M.L. Parry, O.F. Canziani, J.P. Palutikof, P.J. van der Linden, and C.E. Hanson, "Contribution of Working Group II to the Fourth Assessment Report of the Intergovernmental Panel on Climate Change, 2007," Intergovernmental Panel on Climate Change (IPCC) (2007), http://www.ipcc.ch/publications_and_data/ar4/wg2/en/contents.html
72. "FAO/IFAD/WFP Call upon G20 to Redouble Efforts to Fight Hunger," World Food Programme (WFP) (2012), https://www.wfp.org/news/news-release/faoifadwfp-call-upon-g20-redouble-efforts-fight-hunger
73. C. Milmo, "World's Food System Broken, Oxfam Warns," *Independent* (2011), www.independent.co.uk/life-style/food-and-drink/news/worlds-food-system-broken-oxfam-warns-2291469.html
74. "Women: Key to Food Security," Food and Agriculture Organization (FAO) of the United Nations (2011), http://www.fao.org/docrep/014/am719e/am719e00.pdf
75. "Projects in Kenya," World Agroforestry Centre (2017), http://www.worldagroforestry.org/country/kenya/projects

76. "Agroforestry Science to Support the Millennium Development Goals," World Agroforestry Centre (2005), http://www.worldagroforestry.org/downloads/Publications/PDFS/R14568.pdf
77. "Country: Cambodia," System of Rice Intensification (SRI) International Network and Resources Center (2014), http://sri.cals.cornell.edu/countries/cambodia/index.html

Chapter 2

1. "Double Pyramid: A Healthy Diet for All and One Sustainable for the Planet," BCFN (2016), https://www.barillacfn.com/en/dissemination/double_pyramid/
2. Ibid.; I. Contento, *Nutrition Education: Linking Research, Theory, and Practice*, 3rd ed., Jones and Bartlett Learning (2016), http://www.jblearning.com/catalog/9781284078008/
3. C. Matthews, "Livestock a Major Threat to Environment," Food and Agriculture Organization (FAO) of the United Nations (2006), www.fao.org/Newsroom/en/news/2006/1000448/index.html
4. "The BCFN Presents the New Edition of the Double Pyramid," BCFN (2016), http://www.lideamagazine.com/bcfn-foundation-presents-new-version-double-food-environmental-pyramid-usa/
5. Ibid.; BCFN, "Double Pyramid."
6. "Global Report on Diabetes," World Health Organization, Geneva (2016), http://apps.who.int/iris/bitstream/10665/204871/1/9789241565257_eng.pdf; "Diabetes and Cardiovascular Disease," International Diabetes Federation, Brussels, Belgium (2016), www.idf.org/cvd; "Cost of Diabetes Hits 825 Billion Dollars a Year," Harvard T.H. Chan School of Public Health, Boston (2016), https://www.hsph.harvard.edu/news/press-releases/diabetes-cost-825-billion-a-year/
7. T. White and L. Honig, "Corn and Soybean Production Up in 2016, USDA Reports," United States Department of Agriculture, National Agricultural Statistics Service (2017), https://www.nass.usda.gov/Newsroom/2017/01_12_2017.php
8. J.M. Alston, D.A. Sumner, and S.A. Vosti, "Farm Subsidies and Obesity in the United States: National Evidence and International Comparisons," *Food Policy* 33, no. 6 (2008): 470–9, http://www.sciencedirect.com/science/article/pii/S0306919208000523

9. Ibid.
10. S. Fields, "The Fat of the Land: Do Agricultural Subsidies Foster Poor Health?," *Environmental Health Perspectives* (2004), https://www.ncbi.nlm.nih.gov/pmc/articles/PMC1247588/
11. BCFN, "Double Pyramid."
12. "Sources of Greenhouse Gas Emissions," U.S. Environmental Protection Agency (2017), https://www.epa.gov/ghgemissions/sources-greenhouse-gas-emissions#agriculture; "BCFN: Protect the Planet and Our Health with a Mediterranean Diet and Local Products," BCFN Foundation (2016), https://www.barillacfn.com/en/media/bcfn-protect-the-planet-and-our-health-with-a-mediterranean-diet-and-local-products/
13. M. Howe, "Calculating Ecological Footprint Components," Small Farm Permaculture and Sustainable Living (2017), https://small-farm-permaculture-and-sustainable-living.com/ecological_footprint_components/
14. D. Clark, "What's a Carbon Footprint and How Is It Worked Out?" *The Guardian* (4 April 2012), https://www.theguardian.com/environment/2012/apr/04/carbon-footprint-calculated
15. A. Lacis, "CO_2: The Thermostat That Controls Earth's Temperature," National Aeronautics and Space Administration (NASA) Goddard Institute for Space Studies (2010), https://www.giss.nasa.gov/research/briefs/lacis_01/; K. Hansen, "Carbon Dioxide Controls Earth's Temperature," Phys.Org (2010), https://phys.org/news/2010-10-carbon-dioxide-earths-temperature.html#jCp; Clark, "What's a Carbon Footprint."
16. "Climatarian: The 'Zero Emissions' Meal," BCFN Foundation (2016), https://www.barillacfn.com/en/magazine/food-and-sustainability/climatarian-the-zero-emissions-meal/; P. Scarborough, P.N. Appleby, A. Mizdrak, A.D.M. Briggs, R.C. Travis, K.E. Bradbury, and T.J. Key, "Dietary Greenhouse Gas Emissions of Meat-Eaters, Fish-Eaters, Vegetarians and Vegans in the UK," *Climatic Change* 125, no. 2 (2014): 179–92.
17. "Climate Change," *Global Footprint Network: Advancing the Science of Sustainability*, (2017), http://www.footprintnetwork.org/our-work/climate-change/; C. Davenport and M. Landler, "US to Give $3 Billion to Climate Fund to Help Poor Nations, and Spur Rich Ones," *New York Times* (14 November 2014), https://www.nytimes.com/2014/11/15/us/politics/obama-climate-change-fund-3-billion-announcement.html

18. B. Buchner and J. Wilkinson, "The Paris Agreement Is a Signal to Unlock Trillions in Climate Finance," Climate Policy Initiative (2015), https://climatepolicyinitiative.org/2015/12/14/the-paris-agreement-is-a-signal-to-unlock-trillions-in-climate-finance/
19. "The Emissions Gap Report," UNEP (2015), uneplive.unep.org/media/docs/theme/13/EGR_2015_ES_English_Embargoed.pdf; K. Levin and T. Fransen, "Why Are INDC Studies Reaching Different Temperature Estimates?," World Resources Institute (2015), http://www.wri.org/blog/2015/11/insider-why-are-indc-studies-reaching-different-temperature-estimates
20. R. Valentini, "Feed 9 Billion People Respecting the 2°C Limit?," in *Eating Planet: Food and Sustainability: Building Our Future*, BCFN Foundation (2016), https://www.barillacfn.com/media/pdf/Barilla_Eating_planet-2016_Eng_S_abstract.pdf
21. M. Shear, "Trump Will Withdraw US from Paris Climate Agreement," *New York Times* (1 June 2017), https://www.nytimes.com/2017/06/01/climate/trump-paris-climate-agreement.html; N. Prupis, "Obama Sends $500 Million to Green Climate Fund, Signaling End to Era," *Common Dreams* (17 January 2017), https://www.commondreams.org/news/2017/01/17/obama-sends-500-million-green-climate-fund-signaling-end-era
22. "Trump Withdraws U.S. from Paris Agreement; Andrew Cuomo, Barack Obama, Scott Pruitt, More Respond," *Newsday* (2 June 2017), www.newsday.com/news/nation/trump-withdraws-u-s-from-paris-agreement-andrew-cuomo-barack-obama-scott-pruitt-more-respond-1.13700637
23. Ibid.
24. J. Horowitz and S. Strom, "Obama Speaks in Milan, with Food as Text and Politics as Subtext," *New York Times* (9 May 2017), https://www.nytimes.com/2017/05/09/world/europe/obama-food-milan-seeds-chips.html
25. "Biomimicry, Food, and Climate Change: An Interview with Anna Lappé," The Biomimicry Institute, Biomimicry and Climate Change Leaders series (2017), https://biomimicry.org/biomimicry-food-climate-change-interview-anna-lappe/
26. "Regenerative Organic Agriculture and Climate Change: A Down-to-Earth Solution to Global Warming," Rodale Institute (2017), https://rodaleinstitute.org/assets/WhitePaper.pdf

27. Ibid.
28. "What Is a Water Footprint?," Water Footprint Network (2017), http://waterfootprint.org/en/water-footprint/what-is-water-footprint/
29. A. Muhammad-Muaz and M.H. Marlia, "Water Footprint Assessment of Oil Palm in Malaysia: A Preliminary Study," *American Institute of Physics Conference Proceedings* 1618, no. 803 (2014); B. Bras, F. Tejada, J. Yen, J. Zullo, and T. Guldberg, "Quantifying the Life Cycle Water Consumption of a Passenger Vehicle," SAE Technical Paper Series (2012).
30. Ibid.; "Drinking Water Fact Sheet," World Health Organization (WHO) (2017), http://www.who.int/mediacentre/factsheets/fs391/en/
31. Ibid.
32. P. Gleick, N. Ajami, et al. (eds.), *The World's Water, Volume 8: The Biennial Report on Freshwater Resources*, Island Press (2014), https://islandpress.org/book/the-worlds-water-volume-8
33. Ibid.
34. "Drinking-Water," World Health Organization (2017), www.who.int/mediacentre/factsheets/fs391/en/; R.P. Hall, B. Van Koppen, and E. Van Houweling, "The Human Right to Water: The Importance of Domestic and Productive Water Rights." *Science and Engineering Ethics* 20, no. 4 (2014): 849–68, https://www.ncbi.nlm.nih.gov/pmc/articles/PMC4237907/
35. Hall et al., "The Human Right to Water."
36. Ibid.
37. T. Allan, "Virtual Water Between Underconsumption and Poor Management," in *Eating Planet: Food and Sustainability: Building Our Future*, BCFN Foundation (2016), https://www.barillacfn.com/media/pdf/Barilla_Eating_planet-2016_Eng_S_abstract.pdf
38. D. Wichelns, "An Economic Analysis of the Virtual Water Concept in Relation to the Agri-Food Sector," Organisation for Economic Co-operation and Development (OECD) (2010), http://www.fao.org/fsnforum/cfs-hlpe/sites/cfs-hlpe/files/files/Water/Wichelns%20on%20VW.pdf
39. Ibid.
40. D. Michel and A. Pandya, "Troubled Waters: Climate Change, Hydropolitics, and Transboundary Resources," The Henry L. Stimson Center (2009), https://www.globalpolicy.org/images/pdfs/troubled_waters-complete.pdf

41. Ibid.
42. Ibid.
43. S. Varghese, "Privatizing US Water," Institute for Agriculture and Trade Policy (IATP) Trade and Global Governance Program (2015), https://www.iatp.org/files/451_2_99838.pdf
44. Ibid.
45. "Oil and Gas Services: Water Resource Management and Monitoring," EHS Support (2017), http://www.ehs-support.com/services/oil-gas-services/water-resource-management-monitoring
46. J. Specht, "Food Hero Series: Tony Allan, Water Policy Innovator," Food Tank: The Think Tank for Food (2016), https://foodtank.com/news/2013/03/food-hero-series-tony-allan-water-policy-innovator/
47. "Farming: Soil Erosion and Degradation," WWF (2017), http://wwf.panda.org/what_we_do/footprint/agriculture/impacts/soil_erosion/
48. E. Nkonya et al. (eds.), "Economics of Land Degradation and Improvement: A Global Assessment for Sustainable Development," IFPRI (2016), http://www.commonland.com/_doc/5091_699899401.pdf; "Earth Day," The BCFN Foundation and the Milan Center for Food Law and Policy (2017), https://www.barillacfn.com/en/media/earth-day-22nd-april-2017/
49. "Poverty and Climate Change: Reducing the Vulnerability of the Poor through Adaptation," Organisation for Economic Cooperation and Development (OECD) (2003), http://www.oecd.org/env/cc/2502872.pdf
50. "The State of the World's Land and Water Resources: Managing Systems at Risk," Food and Agriculture Organization of the United Nations (2011), www.fao.org/nr/solaw/solaw-home/en/
51. "Earth Day," BCFN Foundation (2017), https://www.barillacfn.com/en/media/earth-day-22nd-april-2017-the-bcfn-foundation-and-the-milan-center-for-food-law-and-policy-each-year-the-planet-loses-an-agricultural-surface-area-as-big-as-the-philippines-in-40-years-30-of-all-arable-land-will-be-infertile-while-hunger-is-already-a-risk/
52. "The State of the World's Land and Water Resources: Managing Systems at Risk," Food and Agriculture Organization (FAO) of the United Nations (2011), http://www.fao.org/docrep/017/i1688e/i1688e.pdf
53. P. Sullivan, "Sustainable Soil Management: Soil Systems Guide," National Sustainable Agriculture Information Service, National

Center for Appropriate Technology (2004), http://cabrillo.edu/~dobrien/soilmgmtATTRA.pdf; D. Pimentel, "Soil Erosion: A Food and Environmental Threat," *Environment, Development and Sustainability* 8, no. 1 (2006), http://saveoursoils.com/userfiles/downloads/1368007451-Soil%20Erosion-David%20Pimentel.pdf; N. Heikkinen, "US Bread Basket Shifts Thanks to Climate Change," *Scientific American* (23 December 2015), https://www.scientificamerican.com/article/u-s-bread-basket-shifts-thanks-to-climate-change/

54. R. Butler, "Soil Erosion and Its Effects," Mongabay (22 July 2012), www.rainforests.mongabay.com/0903.html; T. Echolls, "Soil Erosion due to Rainforest Deforestation," *Sciencing* (25 April 2017), sciencing.com/soil-erosion-due-rainforest-deforestation-23042.html
55. "Savory," *Savory* (2017), www.savory.global
56. "Prolinnova: Promoting Local Innovation," Prolinnova (2017), www.prolinnova.net; "SAVE FOOD: Global Initiative on Food Loss and Waste Reduction," Food and Agriculture Organization (FAO) of the United Nations (2017), http://www.fao.org/save-food/en/
57. "The Library for Food Sovereignty," A Growing Culture (2017), www.agrowingculture.org/library/
58. "Leadership," One Acre Fund (2017), https://oneacrefund.org/about-us/leadership/
59. "SAVE FOOD: Global Initiative on Food Loss and Waste Reduction," Food and Agriculture Organization of the United Nations (2017), www.fao.org/save-food/resources/keyfindings/en/
60. Ibid.
61. L. Furbank, "59 Organizations Fighting Food Loss and Waste," Food Tank (2017), https://foodtank.com/news/2016/07/fighting-food-loss-and-waste/
62. "Introducing Commonsense Bill to Standardize Food Date Labeling," Congresswoman Chellie Pingree, First District of Maine (2016), https://pingree.house.gov/media-center/press-releases/introducing-commonsense-bill-standardaize-food-date-labelng
63. E. Broad Leib, C. Rice, A. Chan, M. Cohen, K. Dimri, M. Malavey, K. Sandson, and D. Trudelle, "Opportunities to Reduce Food Waste in the 2018 Farm Bill," Harvard Food Law and Policy Clinic (2017), www.chlpi.org/wp-content/uploads/2013/12/Opportunities-to-Reduce-Food-Waste-in-the-2018-Farm-Bill_May-2017.pdf

64. L. Furbank, "Food Loss and Waste Protocol Partnership Releases First-Ever Global Food Waste Standard," Food Tank (2017), https://foodtank.com/news/2016/09/food-loss-and-waste-protocol-partnership-releases-first-ever-global-food-wa/
65. Ibid.
66. Ibid.
67. J.N. Pretty et al., "Resource-Conserving Agriculture Increases Yields in Developing Countries," *Environmental Science Technology* 40, no. 4 (2006): 1114–9; "Sustainable Agriculture," National Geographic Society (2015), https://www.nationalgeographic.com/environment/habitats/sustainable-agriculture/; "What Is Sustainable Agriculture?," Agricultural Sustainability Institute at UC Davis (2017), http://asi.ucdavis.edu/programs/sarep/about/what-is-sustainable-agriculture
68. Pretty, "Resource-Conserving Agriculture"; "Sustainable Agriculture," National Geographic Society; "What Is Sustainable Agriculture?," Agricultural Sustainability Institute at UC Davis.
69. "Biodiversity and Agriculture," Healthy and Sustainable Food, Harvard T.H. Chan School of Public Health (2017), www.chgeharvard.org/topic/biodiversity-and-agriculture
70. Ibid.
71. J. Prisco, "Why Bananas as We Know Them Might Go Extinct (Again)," CNN (2016), www.cnn.com/2015/07/22/africa/banana-panama-disease/index.html
72. A. Davis, J. Hill, C. Chase, A. Johanns, and M. Liebman, "Increasing Cropping System Diversity Balances Productivity, Profitability and Environmental Health," *PLoS One* (10 October 2012), www.journals.plos.org/plosone/article?id=10.1371/journal.pone.0047149
73. H. Gould, "10 Things You Need to Know about Sustainable Agriculture," *The Guardian* (1 July 2014), https://www.theguardian.com/sustainable-business/food-blog/sustainable-agriculture-10-things-climate-change
74. "The Future of Food and Agriculture: Trends and Challenges," Food and Agriculture Organization of the United Nations (2017), www.reliefweb.int/report/world/future-food-and-agriculture-trends-and-challenges
75. Gould, "10 Things You Need to Know."
76. Ibid.
77. Ibid.

78. E. Salshutz, "Five Examples of Successful Urban Agriculture Done Differently around the World," Food Tank (2017), https://foodtank.com/news/2013/10/five-different-examples-of-urban-agriculture-from-around-the-world/
79. Ibid.
80. K. Devlin, "Local Foods Offer Tangible Economic Benefits in Some Regions," *Penn State News* (3 February 2014), www.news.psu.edu/story/302490/2014/02/03/research/local-foods-offer-tangible-economic-benefits-some-regions
81. E. Sorensen, "Major Study Documents Benefits of Organic Farming," Washington State University (2014), https://news.wsu.edu/2014/07/11/major-study-documents-benefits-of-organic-farming/
82. O. Balch, "Bread Rationing and Smartcards: Egypt Takes Radical Steps to Tackle Food Waste," *The Guardian* (20 March 2015), https://www.theguardian.com/global-development-professionals-network/2015/mar/20/bread-rationing-egypt-food-waste-grain-wheat
83. F. Allievi, M. Antonelli, and K. Dembska, "The Food Sustainability Index: Fostering the Global Shift towards a More Sustainable Food System," Food Tank (2017). https://foodtank.com/news/2017/05/food-sustainability-index/
84. H. Samuel, "France Enforces Ban on Unlimited Fizzy Drinks in Crackdown on Obesity," *Telegraph* (27 January 2017), www.telegraph.co.uk/news/2017/01/27/france-enacts-ban-unlimited-soda-drinks/
85. Ibid.
86. Balch, "Bread Rationing"; "Blumberg's High-Tech Attempt to Ease Egyptian Grain Drain," Reuters (2016), http://www.reuters.com/article/us-egypt-blumberggrain-idUSKCN0X428Z
87. H. Worley, "Water, Sanitation, Hygiene, and Malnutrition in India," PBR: Population Reference Bureau (2014), http://www.prb.org/Publications/Articles/2014/india-sanitation-malnutrition.aspx
88. Balch, "Bread Rationing."
89. Allievi et al., "The Food Sustainability Index."
90. "The BCFN Reveals the Results of the Food Sustainability Index (FSI)," Barilla Center for Food & Nutrition (2016), https://www.barillacfn.com/en/press_area/the-bcfn-reveals-the-results-of-the-food-sustainability-index-fsi-/

91. As this book was going to press, the Economist Intelligence Unit released an updated version of the Index, ranking a total of 34 countries. "Food Sustainability Index 2017: Global Executive Summary," BCFN and The Economist Intelligence Unit (2017). http://foodsustainability.eiu.com/wp-content/uploads/sites/34/2016/09/FoodSustainabilityIndex2017GlobalExecutiveSummary.pdf
92. "Transforming Cities for a More Sustainable Future," BCFN (n.d.), https://www.barillacfn.com/en/magazine/food-and-sustainability/transforming-cities-for-a-more-sustainable-future/

Chapter 3

1. "Global Processed Food and Beverage Market Outlook, 2017," Frost and Sullivan (2017), https://store.frost.com/global-processed-food-and-beverage-market-outlook-2017.html
2. "A Review of Food Marketing to Children and Adolescents," Federal Trade Commission (2012), https://www.ftc.gov/sites/default/files/documents/reports/review-food-marketing-children-and-adolescents-follow-report/121221foodmarketingreport.pdf
3. "Lobbying Spending: Food and Beverage Industry Profile," Center for Responsive Politics (2017), https://www.opensecrets.org/lobby/indusclient.php?id=N01
4. D. Aaron and M. Siegel, "Sponsorship of National Health Organizations by Two Major Soda Companies," *American Journal of Preventive Medicine* 52, no. 1 (2017), http://www.ajpmonline.org/article/S0749-3797(16)30331-2/fulltext
5. C. Kearns, L. Schmidt, and S. Glantz, "Sugar Industry and Coronary Heart Disease Research," *Journal of the American Medical Association (JAMA) Internal Medicine* 176, no. 11 (2016): 1680–5, http://jamanetwork.com/journals/jamainternalmedicine/article-abstract/2548255
6. M. Burros, "Obamas Bring Their Chicago Chef to the White House," *The New York Times* (28 January 2009), https://thecaucus.blogs.nytimes.com/2009/01/28/obamas-bring-their-chicago-chef-to-the-white-house/
7. R. Swarns, "A White House Chef Wears Two Hats," *The New York Times* (3 November 2009), http://www.nytimes.com/2009/11/04/dining/04kass.html

8. "Sam Kass," Archives: The White House of President Barack Obama (n.d.), https://obamawhitehouse.archives.gov/blog/author/sam-kass
9. S.R. Johnson, "Gauging the Public Health Value of Michelle Obama's Let's Move Campaign," *Modern Healthcare* (2016), www.modernhealthcare.com/article/20160823/NEWS/160829986
10. "Healthy Hunger-Free Kids Act," United States Department of Agriculture (USDA) Food and Nutrition Service (2017), https://www.fns.usda.gov/school-meals/healthy-hunger-free-kids-act
11. S.R. Johnson, "Gauging the Public Health Value of Michelle Obama's Let's Move Campaign," *Modern Healthcare* (2016), http://www.modernhealthcare.com/article/20160823/NEWS/160829986
12. R. Knight and F. Sylla, "Senegal Retail Foods: Retail Food Sector 2012," *USDA Foreign Agriculture Service Global Agricultural Information Network (GAIN) Report* (2013), https://gain.fas.usda.gov/Recent%20GAIN%20Publications/Retail%20Foods_Dakar_Senegal_2-19-2013.pdf
13. M. Seck, "AGRI infos, le nouveau mensuel du monde rural au Sénégal," *Grain de Sel* 37 (2006), www.inter-reseaux.org/IMG/pdf/GdS37-9.pdf
14. Nourishing the Planet, "Changing the World, One Bite at a Time," World Watch (29 October 2010), http://blogs.worldwatch.org/changing-the-world-one-bite-at-a-time
15. Ibid.
16. S. Bailey and D. Nierenberg, "A Global Reason to Eat Locally," *The Vancouver Sun* (17 August 2010), http://www.worldwatch.org/files/pdf/SlowFood_VancouverSun.pdf
17. Kansas State University, "How to Develop Healthy Eating Habits in a Child: Start Early and Eat Your Vegetables," *ScienceDaily* (17 April 2015), www.sciencedaily.com/releases/2015/04/150417103427.htm
18. K. Hesketh, L. Goodfellow, U. Ekelund, A. McMinn, K. Godfrey, H. Inskip, C. Cooper, N. Harvey, and E. van Sluijs, "Activity Levels in Mothers and Their Preschool Children," *Pediatrics* 133, no. 4 (2014), www.pediatrics.aappublications.org/content/133/4/e973.short; S. Couch, "Home Food Environment in Relation to Children's Diet Quality and Weight Status," *Journal of the Academy of Nutrition and Dietetics* 114, no. 10 (2014): 1569–79, https://www.ncbi.nlm.nih.gov/pubmed/25066057

19. "Eating Planet: Food and Sustainability: Building Our Future," BCFN (2016), https://www.barillacfn.com/media/pdf/Barilla_Eating_planet-2016_Eng_S_abstract.pdf
20. Ibid.; Hesketh et al., "Activity Levels in Mothers and Their Preschool Children."
21. "FoodCorps: What You'll Do," FoodCorps (2017), https://foodcorps.org/apply/what-youll-do/
22. P. Koch, R. Wolf, et al., "FoodCorps: Creating Healthy School Environments," The Laurie M. Tisch Center for Food, Education, and Policy at Teachers College, Columbia University (2017), https://foodcorps.org/cms/assets/uploads/2016/06/FoodCorps-Creating-Healthy-School-Environments-Teachers-College.pdf
23. "Eating Planet: Food and Sustainability: Building Our Future," BCFN (2016), https://www.barillacfn.com/media/pdf/Barilla_Eating_planet-2016_Eng_S_abstract.pdf
24. "Dietary Guidelines for Americans: 2015–2020 Eighth Edition," US Department of Health and Human Services and US Department of Agriculture (2015), https://health.gov/dietaryguidelines/2015/resources/2015-2020_Dietary_Guidelines.pdf
25. R.E. Black, C.G. Victora, et al., "Maternal and Child Undernutrition and Overweight in Low-Income and Middle-Income Countries," *Lancet* 382, no. 9890 (2013): 427–51, https://www.ncbi.nlm.nih.gov/pubmed/23746772?access_num=23746772&link_type=MED&dopt=Abstract
26. "Levels and Trends in Child Mortality, Report 2015," United Nations Inter-Agency Group for Child Mortality Estimation (2015), http://www.childmortality.org/files_v20/download/IGME%20Report%202015_9_3%20LR%20Web.pdf
27. "Prevalence of Underweight, Weight for Age," World Health Organization (WHO) (2014), https://data.worldbank.org/indicator/SH.STA.MALN.ZS?locations=IN
28. "National Family Health Survey 4: 2015–16," Government of India, Ministry of Health and Family Welfare (2016), http://rchiips.org/NFHS/pdf/NFHS4/India.pdf
29. L. Haddad, C. Hawkes, E. Udomkesmalee, et al., "Global Nutrition Report 2016: From Promise to Impact, Ending Malnutrition by 2030," Global Nutrition Report Stakeholder Group and International Food

Policy Research Institute (IFPRI) (2016), http://ebrary.ifpri.org/utils/getfile/collection/p15738coll2/id/130354/filename/130565.pdf

30. F. Ngure, B. Reid, J. Humphrey, M. Mbuya, G. Pelto, and R. Stoltzfus, "Water, Sanitation, and Hygiene (WASH), Environmental Enteropathy, Nutrition, and Early Child Development: Making the Links," *Annals of the New York Academy of Sciences* 1308 (2014): 118–28, www.onlinelibrary.wiley.com/doi/10.1111/nyas.12330/full
31. "Nutrition and Students' Academic Performance," Wilder Research (2014), https://www.wilder.org/Wilder-Research/Publications/Studies/Fueling%20Academic%20Performance%20-%20Strategies%20to%20Foster%20Healthy%20Eating%20Among%20Students/Nutrition%20and%20Students%27%20Academic%20Performance.pdf
32. W. Willett, J. Koplan, R. Nugent, C. Dusenbury, P. Puska, and T. Gaziano, "Prevention of Chronic Disease by Means of Diet and Lifestyle Changes," in *Disease Control Priorities in Developing Countries*, 2nd ed., The International Bank for Reconstruction and Development and The World Bank (2006), https://www.ncbi.nlm.nih.gov/books/NBK11795/
33. "Obesity and Overweight: Fact Sheet," World Health Organization (2016), http://www.who.int/mediacentre/factsheets/fs311/en/
34. B. Bush and H. Welsh, "Hidden Hunger: America's Growing Malnutrition Epidemic," *The Guardian* (2015), https://www.theguardian.com/lifeandstyle/2015/feb/10/nutrition-hunger-food-children-vitamins-us
35. H. Ritchie and M. Roser, "Micronutrient Deficiency," Our World in Data (2017), https://ourworldindata.org/micronutrient-deficiency/
36. M. Chan, "Obesity and Diabetes: The Slow Moving Disaster," Director-General's Office of the World Health Organization (WHO) (2016), http://www.who.int/dg/speeches/2016/obesity-diabetes-disaster/en/
37. V. Fulgoni III, D. Keast, R. Bailey, and J. Dwyer, "Food Forticants, and Supplements: Where Do Americans Get Their Nutrients?," *The Journal of Nutrition* 141, no. 10 (2011): 1847–54, http://jn.nutrition.org/content/141/10/1847.full
38. Ibid.; "Nutrition and Students' Academic Performance."
39. "The Experts: What Role Should Government Play in Combatting Obesity?," *The Wall Street Journal* (21 April 2013), https://www.wsj.com/articles/SB10001424127887323741004578419031512580080
40. Ibid.

41. "About Us," Action for Healthy Kids (2017), http://www.actionforhealthykids.org/about-us
42. "The State of Food Insecurity in the World 2015," Food and Agriculture Organization (FAO) of the United Nations (2015), http://www.fao.org/hunger/key-messages/en/
43. "Obesity Update 2017," OECD (2017), http://www.oecd.org/els/health-systems/Obesity-Update-2017.pdf
44. D. Magaña-Lemus, A. Ishdorj, C.P. Rosson III, and J. Lara-Álvarez, "Determinants of Household Food Insecurity in Mexico," *Agricultural and Food Economics* 4, no. 10 (2016), https://agrifoodecon.springeropen.com/articles/10.1186/s40100-016-0054-9
45. "Pobreza y Derechos Sociales de Niñas, Niños, y Adolescentes en México, 2008–2010," Consejo Nacional de Evaluación de la Política de Desarrollo Social (CONEVAL) and UNICEF (2012), https://www.unicef.org/mexico/spanish/UnicefPobreza_web_ene22.pdf
46. "Obesity and Overweight: Fact Sheet," World Health Organization (WHO) (2016), http://www.who.int/mediacentre/factsheets/fs311/en/
47. "Trends in Adult Body-Mass Index in 200 Countries from 1975 to 2014: A Pooled Analysis of 1698 Population-Based Measurement Studies with 19.2 Million Participants," *The Lancet* 387, no. 10026 (2016): 1377–96, http://dx.doi.org/10.1016/S0140-6736(16)30054-X
48. "More Obese People in the World Than Underweight, Says Study," The BBC (2016), http://www.bbc.com/news/health-35933691
49. "Obesity," World Health Organization (WHO) (2014), http://www.wpro.who.int/mediacentre/factsheets/obesity/en/
50. "Three Paradoxes to Solve," BCFN (2017), https://www.barillacfn.com/en/magazine/food-for-all/three-paradoxes-to-solve/
51. K.C. Giacomin and J.O.A. Firmo, "Old Age, Disability and Care in Public Health," *Ciência & Saúde Coletiva* 20, no. 12 (2015): 3631–40, https://dx.doi.org/10.1590/1413-812320152012.11752014
52. W. Leslie and C. Hankey, "Aging, Nutritional Status and Health," *Healthcare* 3, no. 3 (2015): 648–658, http://www.mdpi.com/2227-9032/3/3/648/htm
53. L. Jacobsen, M. Mather, M. Lee, and M. Kent, "America's Aging Population," Population Reference Bureau (2011), http://www.prb.org/pdf11/aging-in-america.pdf

54. A. Kenkmann, G.M. Price, J. Bolton, and L. Hooper, "Health, Wellbeing, and Nutritional Status of Older People Living in UK Care Homes: An Exploratory Evaluation of Changes in Food and Drink Provision," *BMC Geriatrics* 10, no. 28 (2010), https://bmcgeriatr.biomedcentral.com/articles/10.1186/1471-2318-10-28
55. K. Tucker, "Nutrition Concerns for Aging Populations," The Jean Mayer U.S. Department of Agriculture (USDA) Human Research Centre on Aging (HNRCA) (2010), https://www.ncbi.nlm.nih.gov/books/NBK51837/
56. E. Nicklett and A. Kadell, "Fruit and Vegetable Intake among Older Adults: A Scoping Review," *Maturitas* 75, no. 4 (2013): 305–12, https://www.ncbi.nlm.nih.gov/pmc/articles/PMC3713183/
57. M. Osler and M. Schroll, "Diet and Mortality in a Cohort of Elderly People in a North European Community," *International Journal of Epidemiology* 26 (1997): 155–9, https://www.ncbi.nlm.nih.gov/pubmed/9126515; K. Zubair et al., "Life-Years Gained from Population Risk Factor Changes and Modern Cardiology Treatments in Ireland," *European Journal of Public Health* (2006), https://www.ncbi.nlm.nih.gov/pubmed/16798782; M. Hamer et al., "Dietary Patterns, Assessed from a Weighed Food Record, and Survival among Elderly Participants from the United Kingdom," *European Journal of Clinical Nutrition* (2010), https://www.ncbi.nlm.nih.gov/pmc/articles/PMC3398131/; H. Cai et al., "Dietary Patterns and Their Correlates among Middle-Aged and Elderly Chinese Men: A Report from the Shanghai Men's Health Study," *British Journal of Nutrition* (2007), https://www.ncbi.nlm.nih.gov/pubmed/17524168; C.A. Spencer et al., "A Simple Lifestyle Score Predicts Survival in Healthy Elderly Men," *Preventive Medicine* 40, no. 6 (2005): 712–7, https://www.ncbi.nlm.nih.gov/pubmed/15850869
58. J. Jones, M. Duffy, Y. Coull, and H. Wilkinson, "Older People Living in the Community—Nutritional Needs, Barriers and Interventions: A Literature Review," Scottish Government Social Research (2009), http://www.scie-socialcareonline.org.uk/older-people-living-in-the-community-nutritional-needs-barriers-and-interventions-a-literature-review/r/a11G00000017xKZIAY
59. "A Review of Food Marketing to Children and Adolescents," Federal Trade Commission (2012), https://www.ftc.gov/sites/default/files/documents/

reports/review-food-marketing-children-and-adolescents-follow-report/121221foodmarketingreport.pdf

60. "Lobbying Spending: Food and Beverage Industry Profile," Center for Responsive Politics (2017), https://www.opensecrets.org/lobby/indusclient.php?id=N01
61. "The Impact of Food Advertising on Childhood Obesity," American Psychological Association (2017), http://www.apa.org/topics/kids-media/food.aspx
62. D. Kunkel, J. Castonguay, and C. Filer, "Evaluating Industry Self-Regulation of Food Marketing to Children," *American Journal of Preventative Medicine* 49, no. 2 (2015): 181–7, http://www.sciencedirect.com/science/article/pii/S0749379715000951
63. M. Story and S. French, "Food Advertising and Marketing Directed at Children and Adolescents in the US," *International Journal of Behavioral Nutrition and Physical Activity* 1, no. 3 (2004), https://ijbnpa.biomedcentral.com/articles/10.1186/1479-5868-1-3
64. E. Boyland et al., "Advertising as a Cue to Consume: A Systematic Review and Meta-Analysis of the Effects of Acute Exposure to Unhealthy Food and Nonalcoholic Beverage Advertising on Intake in Children and Adults," *The American Journal of Clinical Nutrition* (2016), http://ajcn.nutrition.org/content/early/2016/01/20/ajcn.115.120022
65. R. Boswell and H. Kober, "Food Cue Reactivity and Craving Predict Eating and Weight Gain: A Meta-Analytic Review," *Obesity Reviews* 17, no. 2 (2016): 159–77, https://www.ncbi.nlm.nih.gov/pubmed/26644270
66. J.A. Mennella, "Ontogeny of Taste Preferences: Basic Biology and Implications for Health," *American Journal of Clinical Nutrition* 99, no. 3 (2014), http://ajcn.nutrition.org/content/99/3/704S
67. P. Ueda, L. Tong, C. Viedma, S.J. Chandy, G. Marrone, A. Simon, et al., "Food Marketing Towards Children: Brand Logo Recognition, Food-Related Behavior and BMI among 3–13-year-olds in a South Indian Town," *PLoS One* (2012), http://journals.plos.org/plosone/article?id=10.1371/journal.pone.0047000
68. T.B. Cornwall and A.R. McAlister, "Alternative Thinking about Starting Points of Obesity. Development of Child Taste Preferences," *Appetite* 56, no. 2 (2011): 428–39, http://www.sciencedirect.com/science/article/pii/S0195666311000262?via%3Dihub

69. B. Kelly, S. Vandevijvere, B. Freeman, and G. Jenkin, "New Media but Same Old Tricks: Food Marketing to Children in the Digital Age," *Current Obesity Reports* 4, no. 1 (2015): 37–45, https://link.springer.com/article/10.1007%2Fs13679-014-0128-5
70. A. Lappé, "The Soda Industry's Creepy Youth Campaign," *Real Food Media* (2015), http://realfoodmedia.org/the-soda-industrys-creepy-youth-campaign/
71. M. Nestle, "Food Politics: How the Food Industry Influences Nutrition and Health," University of California Press (2002), https://www.foodpolitics.com/food-politics-how-the-food-industry-influences-nutrition-and-health/
72. Ibid.
73. W. Frazier and J. Harris, "Trends in Television Food Advertising to Young People: 2016 Update," The University of Connecticut Rudd Center for Food Policy and Obesity (2017), http://uconnruddcenter.org/files/TVAdTrends2017.pdf
74. "Local Food with a Purpose," 4P Foods (2017), 4pfoods.com/about-us/
75. P. Holden, "Opening Remarks for Plenary Session on Why Our Food System Must Change," in *The True Cost of American Food 2016 Conference Proceedings* (2016), http://sustainablefoodtrust.org/wp-content/uploads/2013/04/TCAF-report.pdf
76. "Antibiotic Use," Tyson (2017), http://www.tysonfoods.com/media/position-statements/antibiotic-use.aspx
77. A. González, "Largest Organic Grocer Now Costco, Analysts Say," *The Seattle Times* (1 June 2015), http://www.seattletimes.com/business/retail/costco-becomes-largest-organic-grocer-analysts-say/
78. B. Kowitt, "Is the Largest Natural-Foods Brand Even Sold at Whole Foods?," *Fortune* (28 October 2015), http://fortune.com/2015/10/28/kroger-natural-organic-food
79. D.A. Bainbridge, "Rebuilding the American Economy with True Cost Accounting," *Rio Redondo Press* (2009), http://sustainabilityleader.org/Sustainability_Leader/Home_files/bFront%20DedPrefAckTOC.pdf
80. "About the Center for Science in the Public Interest," Center for Science in the Public Interest (2017), https://cspinet.org/about
81. "CSPI Victories," Center for Science in the Public Interest (2016), https://cspinet.org/about/victories

82. "Preliminary Proposed Nutrition Principles to Guide Industry Self-Regulatory Efforts: Request for Comments," Interagency Working Group on Food Marketed to Children (2011), https://www.ftc.gov/sites/default/files/documents/public_events/food-marketed-children-forum-interagency-working-group-proposal/110428foodmarketproposedguide.pdf
83. A. Must, "Eating Planet: Food and Sustainability: Building Our Future," BCFN (2016), https://www.barillacfn.com/media/pdf/Barilla_Eating_planet-2016_Eng_S_abstract.pdf
84. "Zero Hunger," World Food Programme (2017), http://www1.wfp.org/zero-hunger
85. "Overview," World Food Programme (2017), http://www1.wfp.org/overview
86. "About the Program," Africa RISING (2017), https://africa-rising.net/about/
87. "Who We Are," Red de Guardianes de Semillas del Ecuador (2017), http://redsemillas.org/?page_id=217
88. "Agriculture and Food," The Economics of Ecosystems and Biodiversity (2017), http://www.teebweb.org/agriculture-and-food/

Chapter 4

1. C. Lévi-Strauss, *The Raw and the Cooked* (New York: Harper & Row, 1969); M. Pollan, *Cooked: A Natural History of Transformation* (New York: The Penguin Press, 2013).
2. "The Cultural Dimension of Food," BCFN (2016), https://www.unscn.org/web/archives_resources/files/the_cultural_dimension_of_food.pdf
3. N. Rayman, "How a McDonald's Restaurant Spawned the Slow Food Movement," *Time*, (2014), www.time.com/3626290/mcdonalds-slow-food/
4. N. Rayman, "How a McDonald's Restaurant Spawned the Slow Food Movement," *Time* (10 December 2014).
5. "What Is Happening to Agrobiodiversity?," Food and Agriculture Organization (FAO) of the United Nations (1999), www.fao.org/docrep/007/y5609e/y5609e02.htm
6. L. Govender et al., "Food and Nutrition Insecurity in Selected Rural Communities of KwaZulu-Natal, South Africa: Linking Human Nutrition and Agriculture," *International Journal of Environmental Research and Public*

Health, 14, no. 1 (2017): 17, https://www.ncbi.nlm.nih.gov/pmc/articles/PMC5295268/

7. A.D. Jones and G. Ejeta, "A New Global Agenda for Nutrition and Health: The Importance of Agriculture and Food Systems," *Bulletin of the World Health Organization* 94 (2015): 228–9, http://www.who.int/bulletin/volumes/94/3/15-164509/en/
8. "Obesity and Overweight," World Health Organization (2016), www.who.int/mediacentre/factsheets/fs311/en/
9. "Slow Food USA," Slow Food (2017), https://www.slowfoodusa.org
10. "Ark of Taste," Slow Food (2017), https://www.slowfoodusa.org/ark-of-taste
11. H. Heuler, "Ugandan Promotes 'Slow Food' Ideals to Feed Africa," *VOA News* (24 July 2014), https://www.voanews.com/a/ugandan-promotes-slow-food-ideals-to-feed-africa/1964162.html
12. "10,000 Gardens in Africa," Slow Food Foundation for Biodiversity (2017), https://www.fondazioneslowfood.com/en/what-we-do/10-000-gardens-in-africa/
13. S. Small, "34 Indigenous Crops Promoting Health and Feeding the World," Food Tank (2017), https://foodtank.com/news/2015/08/indigenous-crops-promoting-health-feeding-world/
14. Ibid.
15. "Mandacaia Bee Honey (*Melipona mandacaia*)," Slow Food Foundation for Biodiversity (2017), https://www.fondazioneslowfood.com/en/ark-of-taste-slow-food/mandacaia-bee-melipona-mandacaia-honey/
16. Ibid.
17. "Caatinga Mandacaia Bee Honey," Slow Food Foundation for Biodiversity (2017), https://www.fondazioneslowfood.com/en/slow-food-presidia/mandacaia-bee-caatinga-honey/
18. "Maungo," Slow Food Foundation for Biodiversity (2017), https://www.fondazioneslowfood.com/en/ark-of-taste-slow-food/maungo-5/
19. Ibid.
20. L. Estes and L. Greyling, "Southeastern Africa: South Africa, Mozambique, Botswana, Zambia, Zimbabwe, Swaziland, Namibia, and Malawi," WWF (2017), https://www.worldwildlife.org/ecoregions/at0725
21. "5 Endangered North American Plants with Significant Medicinal Benefits," Herbs List (2012), www.herbslist.net/5-endangered-north-american-plants-with-significant-medicinal-benefits.html

22. Ibid.
23. "Conservation and Restoration of North American Ginseng," Espace pour la Vie Montreal (2017), www.espacepourlavie.ca/en/conservation-and-restoration-north-american-ginseng
24. "Nature, Culture, and History: Refuge in the Mountains," Queensland Government Department of National Parks, Sport, and Racing (2013), https://www.npsr.qld.gov.au/parks/bunya-mountains/culture.html
25. Ibid.
26. Ibid.
27. "Millets, Sorghum, and Grain Legumes: The Smart Foods of the Future," International Crops Research Institute for the Semi-Arid Tropics (2017), www.icrisat.org/millets-sorghum-and-grain-legumes-the-smart-foods-of-the-future/
28. Ibid.
29. Ibid.
30. "Guinea Hog," Slow Food USA (2017), https://www.slowfoodusa.org/ark-item/guinea-hog
31. Ibid.
32. "Grant Jumpstarts Largest Heritage-Hog Conservation Effort," Hobby Farms (2012), www.hobbyfarms.com/grant-jumpstarts-largest-heritage-hog-conservation-effort-2/
33. "Holland Chicken," The Livestock Conservancy (2017), https://livestockconservancy.org/index.php/heritage/internal/holland
34. Ibid.
35. S. Weaver, "Raising Rare Livestock Breeds," Hobby Farms (2009), www.hobbyfarms.com/raising-rare-livestock-breeds-7/
36. "Abouriou Wine," Wine-Searcher (2017), https://www.wine-searcher.com/grape-1968-abouriou
37. "Abouriou Grapes," Slow Food (2017), https://www.fondazioneslowfood.com/en/ark-of-taste-slow-food/abouriou-grape-variety/
38. Stevie Stacionis, "Abouriou: An Endangered Species of Wine," Serious Eats (2017), http://drinks.seriouseats.com/2013/02/adventures-with-weird-wine-grapes-abouriou-california-endangered-species-wine.html
39. S. Fields, "The Fat of the Land: Do Agricultural Subsidies Foster Poor Health?," *Environmental Health Perspectives* 112, no. 14 (2004).
40. V. Shiva, "From Seeds of Suicide to Seeds of Hope: Why Are Indian Farmers Committing Suicide and How Can We Stop This Tragedy?,"

Huffington Post (2017), www.huffingtonpost.com/vandana-shiva/from-seeds-of-suicide-to_b_192419.html

41. D. Moss, "Why a Public Sector Indian Agronomist Embraced Agroecology," Food Tank (2017), https://foodtank.com/news/2017/04/indian-indigenous-agroecological-traditions/
42. "Participatory Guarantee Systems (PGS)," IFOAM Organics International (2017), www.ifoam.bio/en/organic-policy-guarantee/participatory-guarantee-systems-pgs
43. World Vegetable Center (2017), https://avrdc.org
44. "Managing Germplasm," World Vegetable Center (2017), https://avrdc.org/our-work/managing-germplasm/
45. Ibid.
46. "Smart Food," International Crops Research Institute for the Semi-Arid Tropics (ICRISAT) (2017), www.icrisat.org/smartfood/
47. Ibid.
48. "Millets and Sorghum: Climate-Smart Grains for a Warmer World," CGIAR Development Dialogues 2014 (2014), www.dialogues.cgiar.org/blog/millets-sorghum-climate-smart-grains-warmer-world/
49. Ibid.
50. G. Sertorio and M.C. Martinengo, "Consumare. Lineamenti di sociologia dei consume," G. Giappichelli Editore (2005), https://www.ibs.it/consumare-lineamenti-di-sociologia-dei-libro-guido-sertorio-m-cristina-martinengo/e/9788834857182
51. Ibid.
52. Ibid.
53. "The Truth about Land Grabs," Oxfam (2017), https://www.oxfamamerica.org/take-action/campaign/food-farming-and-hunger/land-grabs/
54. "The State of Food and Agriculture: Women in Agriculture, Closing the Gender Gap for Development," Food and Agriculture Organization (FAO) of the United Nations (2011), www.fao.org/docrep/013/i2050e/i2050e.pdf
55. Ibid.
56. "Lindsey Shute: 'Agriculture Is the Wealth of the Nation,'" Food Tank (2017), https://foodtank.com/news/2017/03/lindsey-shute-interview/
57. A. Trichopoulou et al., "Definitions and Potential Health Benefits of the Mediterranean Diet: Views from Experts around the World," *BMC*

Med 12, no. 112 (2014), https://www.ncbi.nlm.nih.gov/pmc/articles/PMC4222885/

58. Ibid.
59. "Eating Planet: Food and Sustainability: Building Our Future," BCFN (2016), https://www.barillacfn.com/media/pdf/Barilla_Eating_planet-2016_Eng_S_abstract.pdf
60. A. Brillat-Savarin, "The Physiology of Taste, or Transcendental Gastronomy" (1826), http://www.gutenberg.org/ebooks/5434
61. Ibid.
62. Ibid.; Sertorio and Martinengo, "Consumare: Lineamenti di sociologia dei consume"; H. Marvin, "Good to Eat: Riddles of Food and Culture," Waveland Press (1998), https://www.waveland.com/browse.php?t=158
63. Sertorio and Martinengo, "Consumare: Lineamenti di sociologia dei consume"; Marvin, "Good to Eat."
64. D.M. Douglas and B. Isherwood, "The World of Goods: Towards an Anthropology of Consumption," Routledge (1996).
65. "BCFN: Protect the Planet and Our Health with a Mediterranean Diet and Local Products," BCFN (2016), https://www.barillacfn.com/en/media/bcfn-protect-the-planet-and-our-health-with-a-mediterranean-diet-and-local-products/
66. Ibid.
67. Ibid.
68. C. Fischler, "With Food for Free We Can Cancel Our Identity," Corriere della Sera–Sette (2012), http://manuelamimosaravasio.com/claude-fischler-cibo/
69. D.L. Katz and S. Meller, "Can We Say What Diet Is Best for Health?," *Annual Review of Public Health* 35 (2014): 83–103, http://www.annualreviews.org/doi/full/10.1146/annurev-publhealth-032013-182351
70. "Food and the Environment," BCFN (2013), https://www.barillacfn.com/m/publications/bcfn-magazine-foodenvironment1.pdf
71. A. Rosenberger, "Five Questions with Tony Hillery, Founder of Harlem Grown," Food Tank (2016), https://foodtank.com/news/2016/08/five-questions-with-tony-hillery-founder-of-harlem-grown/; Jamie's Food Revolution (2017), www.jamiesfoodrevolution.org
72. M. Hammershoj, U. Kidmose, and S. Steenfeldt, "Deposition of Carotenoids in Egg Yolk by Short-Term Supplement of Coloured Carrot

(*Daucus carota*) Varieties as Forage Material for Egg-Laying Hens," *Journal of the Science of Food and Agriculture* 90, no. 7 (2010): 1163–71, https://www.ncbi.nlm.nih.gov/pubmed/20393997

73. J.K. Boehm, D.R. Williams, E.B. Rimm, C. Ryff, and L.D. Kubzansky, "Association between Optimism and Serum Antioxidants in the Midlife in the United States Study," *Psychosomatic Medicine* 75, no. 1 (2013): 2–10, https://www.ncbi.nlm.nih.gov/pubmed/23257932
74. D. Morgan et al., "Seattle Food System Enhancement Project: Greenhouse Gas Emissions Study," University of Washington Program on the Environment (2008), https://my.vanderbilt.edu/danmorgan/files/2012/12/Final_GHG_Report1.pdf
75. K. Musick and A. Meier, "Assessing Causality and Persistence in Associations between Family Dinners and Adolescent Well-Being," *Journal of Marriage and Family* 74, no. 3 (2012): 476–93, www.onlinelibrary.wiley.com/doi/10.1111/j.1741-3737.2012.00973.x/abstract
76. Ibid.
77. M. Franchi, "Il cibo flessibile: Nuovi comportamenti di consume," Carocci Editore (2009), http://www.carocci.it/index.php?option=com_carocci&task=schedalibro&Itemid=72&isbn=9788843049370
78. "State of the World 2011: Innovations That Nourish the Planet," *Worldwatch Institute: Vision for a Sustainable World* 10–11 (2011), www.worldwatch.org/sow11
79. "Outlook on the Millennial Consumer 2014," The Hartman Group (2014), www.store.hartman-group.com/content/millennials-2014-overview.pdf
80. "Americans' Budget Shares Devoted to Food Have Flattened in Recent Years," USDA (2014), https://www.ers.usda.gov/data-products/chart-gallery/gallery/chart-detail/?chartId=76967
81. "Kernza Grain: Toward a Perennial Agriculture," The Land Institute (2017), https://landinstitute.org/our-work/perennial-crops/kernza/
82. A. Savory, "How to Fight Desertification and Reverse Climate Change," TED Talk (2013), https://www.ted.com/speakers/allan_savory
83. SEWA: Self-Employed Women's Association (2009), www.sewa.org
84. N. Nunn and N. Qian, "Aiding Conflict: The Impact of U.S. Food Aid on Civil War," *American Economic Review* 104:6 (2014), https://www.aeaweb.org/articles?id=10.1257/aer.104.6.1630; C. Dugger, "Charity

Finds That U.S. Food Aid for Africa Hurts Instead of Helps," *The New York Times* (14 August 2007), http://www.nytimes.com/2007/08/14/world/americas/14iht-food.4.7116855.html

85. "Eating in 2030: Trends and Perspectives," BCFN (2012); CoHo Ecovillage (2017), https://www.cohoecovillage.org; "Shared Meals," Community Learning Incubator (2017), www.clips.gen-europe.org/shared-meals/
86. R. Derousseau, "Financial Supper Clubs and Gaming for Millennial Investors," *USA Today* (27 January 2015), https://www.usatoday.com/story/money/personalfinance/2015/01/27/ozy-investment-education-geared-toward-millennials/22402783/
87. "School Meals," World Food Programme (2017), www.1.wfp.org/school-meals

Index

Figures/photos/illustrations are indicated by an "f" and tables by a "t."

Island Press | Board of Directors

Pamela B. Murphy
(Chair)

Terry Gamble Boyer
(Vice-Chair)
Author

Deborah Wiley
(Secretary)
Chair
Wiley Foundation, Inc.

Tony Everett
(Treasurer)

Decker Anstrom
Board of Directors
Discovery Communications

Melissa Shackleton Dann
Managing Director
Endurance Consulting

Katie Dolan
Environmental Writer

Margot Ernst

Alison Greenberg
Executive Director
Georgetown Heritage

David Miller
President
Island Press

Georgia Nassikas
Artist

Alison Sant
Cofounder and Partner
Studio for Urban Projects

Ron Sims
Former Deputy Secretary
US Department of Housing
and Urban Development

Sandra E. Taylor
Chief Executive Officer
Sustainable Business
International LLC